Soldering

BEYOND the BASICS

Techniques to build confidence and control

Joe Silvera

KALMBACH BOOKS

Kalmbach Books
21027 Crossroads Circle
Waukesha, Wisconsin 53186
www.Kalmbach.com/Books

Published in 2014
18 17 16 15 14   1 2 3 4 5

Manufactured in the United States of America

ISBN: 978-1-62700-029-1
EISBN: 978-1-62700-030-7

Editor: Mary Wohlgemuth
Art Director: Carole Ross
Technical Editor: Annie Pennington
Photographers: James Forbes, William Zuback
Illustrator: Kellie Jaeger

Publisher's Cataloging-in-Publication Data

Silvera, Joe, author.
   Soldering beyond the basics : techniques to build confidence and control / Joe Silvera.

      pages : color illustrations ; cm

   Includes index.
   Issued also as an ebook.
   Companion DVD also available from publisher.
   ISBN: 978-1-62700-029-1

   1. Solder and soldering—Handbooks, manuals, etc. 2. Metal-work—Handbooks, manuals, etc. 3. Jewelry making—Handbooks, manuals, etc. I. Title.

TT212 .S552 2014
739.274

CONTENTS

PRACTICE MAKES PERFECT

I recently learned something interesting about skills. Most of us advance in a skill to a certain point and stop. For example, it's hard to advance past your best typing speed. Once this plateau is reached, after struggling to learn the basics, getting better and better with practice, we're able to type without having to focus on it, and our mind is free to wander as we work. But without returning to the struggle, to having to think hard about your practice and how to advance, we stay content at that level. We can define geniuses, artists, and prodigies as people with the fortitude and discipline to push themselves back to the beginning in order to move beyond their limits. Fortunately, soldering is a craft that doesn't encourage resting on your laurels. Every project presents a new challenge, something different to tackle, so that you really do get better with time and practice.

My first book, *Soldering Made Simple: Easy Techniques for Kitchen Table Jewelers*, was written for my students and for anyone who wanted to learn how to solder, whether they had some background in jewelry or not. It thoroughly explains the basics of soldering and lays out a course of guided projects from simple earrings to bezel-set rings. The first book, still in print with thousands of readers worldwide, provides a great introduction to soldering with home-friendly tools.

This book is for those of you who have reached that skill level plateau and want to move beyond the basics, to ascend to more-interesting projects, including more stone setting, soldering near stones, and working with gold and mixed metals. Some explanations of basic techniques and

"Every project presents a new challenge, something different to tackle, so that you really do get better with time and practice."

tools overlap with *Soldering Made Simple*. And I've included a few basic projects to help any beginners with the impatient courage to dive right in to more intermediate designs. But I've filled these pages with lots of juicy tips to help you make more-advanced jewelry, which involves beyond-the-basics soldering techniques, such as using two butane torches or small oxy/propane torches, firescale-retardant fluxes, and anti-flux.

Part 1 reviews materials, tools, setting up a studio, and soldering and stone-setting techniques, including the intermediate skills required for the projects. **Part 2** is made up of 16 projects to practice soldering bezels, prongs, and tube and flush settings; hollow forms and sculpted pendants; protecting stones with heat shields; precision-soldering a gold ring; and working with sensitive gold-filled metal. In **Part 3**, a basics review in the back of the book, you can revisit sawing, filing, and other fundamental metalwork techniques.

Just as in my first book and DVD, I'm committed to using tools and equipment that are safe for a small shop or home studio. All it takes is a corner of a room to set up a simple jewelry bench. But that corner can overflow with hours of satisfied creation as you master the art of soldering.

1 Getting Started

Materials

WHAT CAN BE SOLDERED?

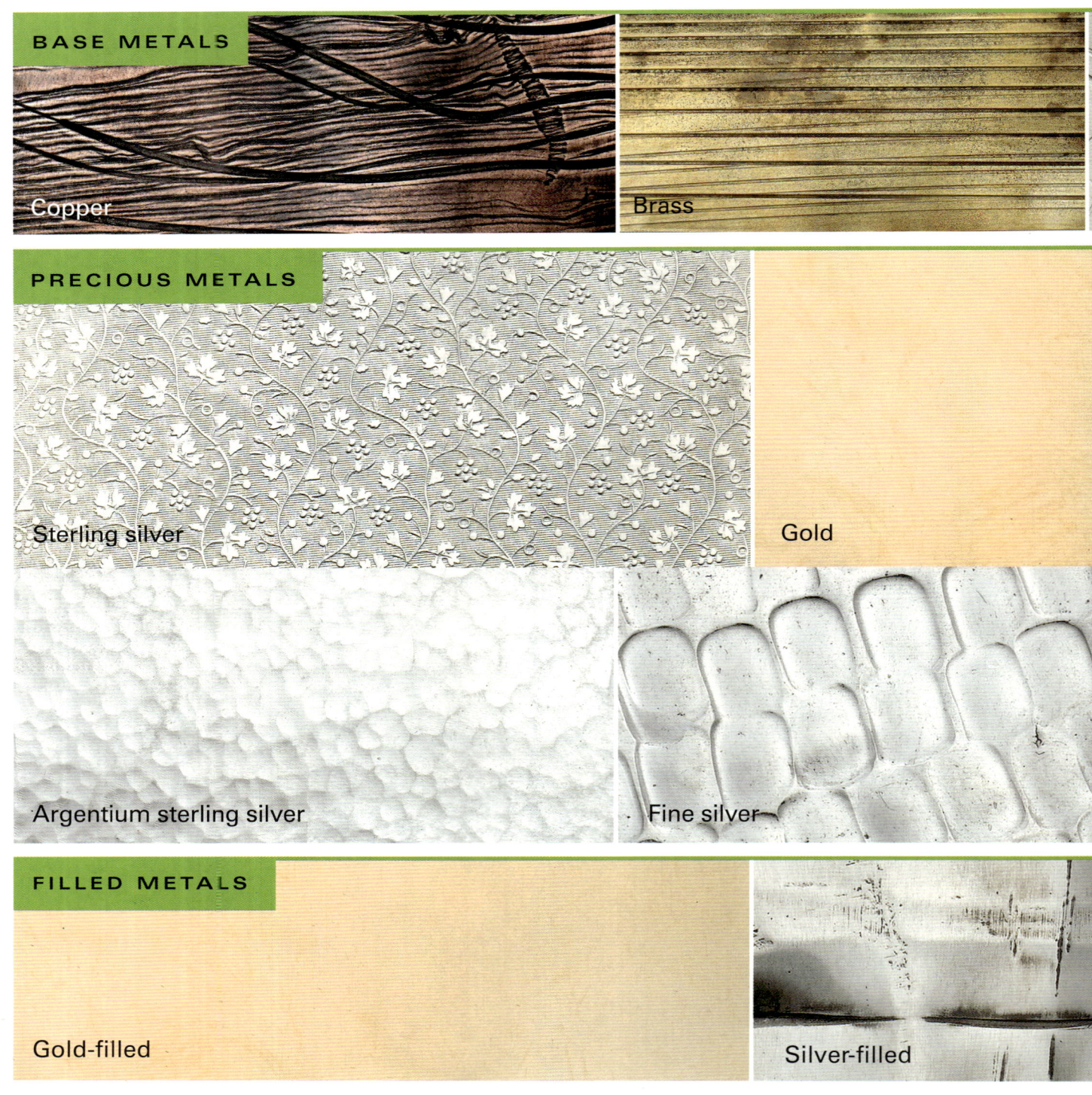

BASE METALS

Copper

Brass

PRECIOUS METALS

Sterling silver

Gold

Argentium sterling silver

Fine silver

FILLED METALS

Gold-filled

Silver-filled

Nickel

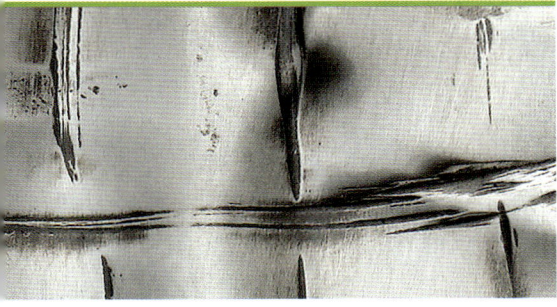

Precious metals and base metals

Jewelry created by studio artists can be made of a lot of materials, but the metals are usually precious metals or base metals, or a combination of both. These metals can be soldered. They're nonferrous, meaning they don't contain any iron. They can be pure metals, sometimes called fine or .999, which refers to the percentage or parts of pure metal they contain. In this case, they are 99.9% pure. They can also be alloys, which are mixtures of two or more pure metals, melted together to make different colors or working properties. The alloy number again refers to the percentage of the dominant metal in the alloy, so sterling silver is .925 or 92.5% fine silver and .075 or 7.5% copper. Filled metals and bimetals, which are made of two metals bonded together, can be soldered too. But unlike regular precious and base metal alloys, they can't be melted or the metals will mix into a weird alloy of base and precious metal.

What can't be soldered?

Soldering is used to join metals, so it doesn't work on other materials. Plated metals don't solder very well. They can be soldered, but the plating will burn off, exposing the base metal core. This includes copper-core craft wire, vermeil, and silver- and gold-plated metal. Base metal findings and wire can also be sealed in tarnish-resistant coatings that burn during soldering, making them too dirty to solder.

White metals, such as pewter, bismuth, tin, and antimony, are low-temp metals that melt below 1200°F (649°C) or even 600°F (316°C). They can't be soldered with hard solders used for silver and gold. They should not be mixed with base or precious metals; they can contaminate the surface.

Gold-filled & silver-filled

Filled metals are metals that are made of two metals bonded together, usually a precious metal to a base metal, to help make the price more economical for fabrication. Silver-filled is sterling or fine silver over red brass. Gold-filled is 12k or 14k (karat) gold bonded to brass or a similar base metal. The advantage of filled metal over plated metal is that the layer of precious metal is thicker than plating, allowing for fabrication, like soldering and some polishing.

1/10 or 1/20

Filled metals are sold and marked by their weight of precious metal: 1/10th is 10% and 1/20th is 5%. Gold-filled is marked and sold by its karat and the weight of that karat. For example, 12KY/20 gold-filled wire is 1/20th or 5% yellow gold by weight. It has a tube of 12k yellow gold around a core of brass. 14KY/20 is the same as 12KY/20, but the gold is 14k. Filled metals should always be hallmarked appropriately. It is not ethical to mark 14KY/20 as 14KY. Instead it should be marked 14K/20, or 14KY/20 GF, or 14K GF, or GF (gold-filled).

Single-clad or double clad

Filled sheet metal can be single sided or double sided. For example, single-clad silver-filled sheet has a sterling side bonded to a brass base. Double-clad has a layer of precious metal on both sides of a brass sheet. Double-clad is more expensive, but it looks like gold, for example, on both sides. Of course, the edges expose the brass core and will need to be covered if you want to hide that. Filled findings like bezels and bezel wire are available. These will often be double clad to hide any base metal cores.

Soldering Tools

The micro butane torch (front), the large-flame butane torch, and a fuel canister.

My tank holder can be wheeled around or fixed in place.

Tank torch tips can be changed to fit the job at hand.

Torches

You have to have a torch to solder, but it doesn't have to be a big tank torch. It's easy to solder with friendly butane cooking torches, or to set up a small-torch system that runs on small tanks—even disposable ones!

Butane torches Butane torches come in a few sizes: pencil, micro butane, and large flame or jumbo. The pencil size has a tiny flame, so I don't recommend it. Micro butane torches are comfortable to hold and can handle soldering jobs up to small bezel settings. For bigger jobs, use a large-flame butane torch. Read and follow the instructions that come with your torch to safely light and refuel your torch.

Tank torches A tank torch runs on a single fuel, like propane or acetylene, or a mix of two gases, like oxygen and propane or oxygen and acetylene. Their flames are generally hotter than butane for faster and more efficient soldering. I prefer the versatility of small torches that run on clean oxy/propane. Their tips can be interchanged for tiny, pencil-point flames up to a melting tip for the largest flame. See *Setting Up a Studio: Setting up a small torch system*, p. 17, for examples of different systems as well as advice on how to set them up.

How to refuel your butane torch

Butane is available in canisters at hardware stores, smoke shops, or convenience stores. Match the nozzle on the canister to the fuel port on the bottom of the torch. Wear safety glasses and refuel your torch away from open flames.

Make sure your torch is off and the flame is out. Find the gas control and turn it off (or all the way toward the minus sign).

Hold the torch upside-down. Remove the stand. Shake the gas canister a few times to warm the fuel up. Hold it away from your face, with the nozzle away from your skin. Hold the butane canister upside down, aligned with the fuel port.

Press down hard with the canister to make a tight seal and to start fueling. If you don't press hard enough, or the nozzle is misaligned, or the torch is full, the butane will spray back right away. Normally, if the torch is partially empty, the butane will hiss into the torch quietly with a slight haze in the air. This can take 5–20 seconds. As soon as the torch is full, wet butane will spray into the air. Stop fueling. Don't overfill your torch! If you do, butane will spit out of the bottom or top and you can't use your torch until this stops.

Turn your torch upright and wait 5–10 minutes after fueling before you use it. This will help to settle any air bubbles in the butane.

Press down hard to make a tight seal for refueling.

Disposable fuel and oxygen tanks are available at hardware stores.

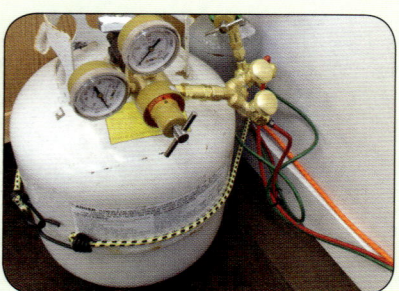

Big propane, acetylene, and oxygen tanks can be refilled at welding suppliers.

Charcoal, honeycomb, and firebrick

Materials like compressed charcoal, honeycomb, and firebrick amplify the heat from your torch and speed up melting and soldering. Charcoal makes a reducing atmosphere when heated, which fights oxides like firescale. Soft charcoal is good for pressing in props or pieces to hold them during soldering. It can smolder if you don't quench it. Hard charcoal will hold heat, doesn't burn away, and lasts longer than soft charcoal. Honeycomb has a clean ceramic surface and is easier to clean. Soft firebrick, available from ceramics suppliers, is very reflective of heat, and can be cut into smaller pieces with a wood saw.

Solder boards

Solder boards are made to take the heat of soldering and come in different materials and brands, like ceramic, Solderite, Transite, and Silquar. The difference between them is preference and how cold the board is to work on. Cold boards like ceramic or Silquar take more heat and can slow soldering down, which can be a good thing when you're working on small jump rings or wire that could be easy to melt. Solderite reflects more heat and can amplify your torch, but pits can be melted into the surface.

Flux

Flux keeps metal clean and helps solder to flow. There are many different types of flux, but they can be grouped into a couple of categories: flux, fire coats, and combinations of both (see *All About Flux*, p. 22).

Soldering picks and tweezers

Soldering picks and tweezers are used to move solder and hot metal. Solder won't melt onto tungsten and titanium picks. Tweezers come in two general groups: locking and non-locking. Non-locking tweezers are easier to use for picking up small things, like jump rings. Cross-locking tweezers are great for holding metals during soldering. Upgrade to tweezers with fiberboard handles that protect your fingers from the heat.

Cleaning your soldering tools

Clean your solder board with a damp towel after it's cooled. Don't use any chemical cleaners. Honeycomb, steel mesh, and solder boards can be cleaned in a utility sink or bucket by pouring boiling hot water over them to dissolve any baked-on flux. Charcoal and firebrick can be sanded with 60-grit sandpaper. Rinse unheated flux from your tools with water. Remove solder from your pick and tweezers with an inexpensive file. Remove hardened flux by filing or pickling the tips of your pick or tweezers in citric pickle for 20 seconds. Rinse and wipe with a towel.

These surfaces amplify heat and speed up melting and soldering.

Solder boards have pros and cons, so you may want a variety on hand to choose the right one for the job.

Left to right: basic paste flux, self-pickling flux, and Cupronil, which is a spray-on firescale-preventative flux.

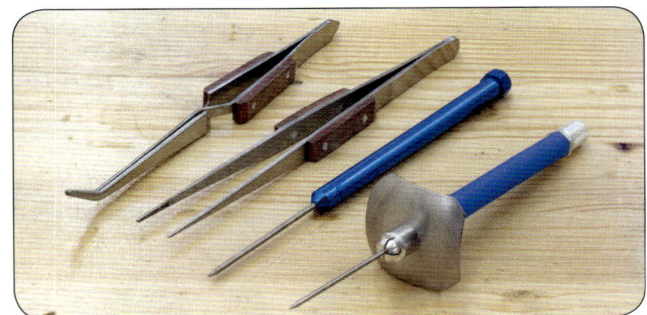

Left to right: cross-locking tweezers, non-locking tweezers, titanium pick, and a shielded tungsten pick.

Jewelry Tools

Left to right: rotary tool with optional handpiece, flex shaft, bench-top polisher.

To shape wire and metal before soldering, use a set of jewelry pliers, flat/half-round forming pliers, and flush-cutters.

Rotary tools, flex shafts, and bench-top polishers

Rotary tools (Dremel is a well-known brand), flex shafts, and bench-top polishers are used for polishing by hand. The biggest difference between them is the strength of their motors at low speed for finishing and setting. Dremels have less torque than a flex shaft, and the tool may stop completely at low speed.

A flex shaft, which you'll find on most professional jewelers' benches, is my preferred power tool. It has a pedal for easy speed control; just turn it on and off with your foot. Its versatile chuck can fit tiny drill bits on up to various sizes of polishing mandrels. A Dremel doesn't always include a smaller, more comfortable flex shaft handle, but they're available as an accessory. A bench-top polisher can be mounted with larger wheels for faster polishing. Use them with 2–3" (51–76mm) 6-ply radial bristle disks, since they don't require any kind of ventilation because there's no dust! Read and follow the instructions that come with your power tools.

Abrasives ranging from micron-graded polishing paper to a jeweler's saw help you shape and refine metal before soldering.

Forge metal with a variety of steel hammers, a bench block, and mandrels in various shapes. A rawhide mallet is useful for shaping.

Pliers

You'll use a variety of jewelry pliers to open and close jump rings and to shape your metal. Buy an inexpensive set of roundnose, flatnose, chainnose, and flush-cutting pliers, because soldering can damage expensive beadwork pliers. Don't use pliers to hold hot metal, to hold work for fluxing, or to quench, because it will ruin them; use tweezers instead. Flat/half-round steel forming pliers are useful for closing rings and shaping bezels.

Saws, files, and sandpaper

A jeweler's saw frame holds delicate but powerful blades that can cut through mild steel. A common size has a 3–4" (76–102mm) throat for sawing deep into sheet metal. Affordable saw blades are sold by the gross in different grades of coarseness, from 4 rough to 8/0 ultra-fine. I recommend sizes 2, 2/0, and 4/0 for most sawing jobs.

Files come in different sizes, shapes, and cuts, from coarse to fine. Hand files are larger, with 6–8" (15.2–20.3cm) cutting lengths. Needle files are small and thin, and mini files are even finer. All files are more comfortable to use with handles. Useful shapes for files include half-round, flat or pillar, tri-square, square, round, and barrette. The higher the number of the cut, the finer the file. I like to have flat and half-round files in each of the three sizes, in cuts 0, 2, and 4. Use rough 0 and medium 2 files to quickly shape and refine your piece. Remove file marks with fine-cut 4 files.

Hammers, anvils, and mandrels

You'll have a lot of choice in hammers for jewelry and metalsmithing. A chasing hammer is made for hitting punches and stamps with its wide face, and the ball end can be used for texturing and shaping metal. Another useful shape is the goldsmithing hammer, with a polished flat face for making crisp flat sides, and a narrow wedge for texture and forging. Consider also getting a planishing hammer with a polished, curved face for forging, and a plastic side for shaping metal without creating any texture. Polished hammers have a mirror finish and will burnish your metal as you work. Rawhide leather mallets will flatten and shape metal without texture. If you like texture, texture hammers have patterns cut in their faces and will quickly imprint your metal.

All of these tools require a steel block or anvil to support the metal as you hammer. The block should also have a smooth finish, to avoid accidental work marks. Place it on a leather mat or mouse pad to absorb a lot of the noise. Some blocks come with a nylon block that can be used to flatten metal or as a softer anvil when texturing two sides.

Steel mandrels are used to shape your metal. Ring mandrels should be steel with fine grooves to mark the sizes. Oval bracelet mandrels can be wood or steel, and if stepped, make it easier to make a cuff. Small bezel mandrels are useful for reshaping bezels and jump rings.

Setting Up a Studio

Working at home

My students ask me all the time about setting up a studio. Most of us learn jewelry in a classroom, where the metalsmithing department is full of benches, tools, and polishing and casting equipment. Translating that big space into your own studio is overwhelming, but the truth is that a jewelry studio can be as small as one table or a bench.

Now, the words "home" and "studio" may not seem like a reasonable combination, especially when considering using jewelry tools, torches, flex shafts, and such. The topic brings up common questions, like: Is it safe? Where should I put my studio? Can I solder at home? The answers are yes, almost anywhere, and absolutely. I've worked in goldsmith shops and lots of studios. They were parts of retail shops, home studios, or office spaces. Floors ranged from wood to carpeting—not your first choices in fire safety. Bench tops were made of wood walls were normal drywall. And yet, we never had a fire. Why? Because we practiced commonsense safety rules.

Later, when I became an independent jeweler, I took what I learned about workplace safety to my home, where my studios were spare bedrooms, half of a room, a shed in the yard, and even a tiny space outside of my bathroom! That space contained my bench, a casting machine, polishing motor, kiln, and everything else used to solder, fabricate, and cast jewelry. With a bit of organization, your studio can start with just a bench or sturdy table. And once it's set up, you can make jewelry anytime you want—as long as you don't wake up the neighbors!

Set up for safety

When soldering, your number-one concern is safety. Sawing, filing, and even polishing with the clean bits I recommend are benign operations when compared to flames and red-hot metal. With a few simple precautions, you can solder safely at home. For more tips, read *Safety first*, p. 19.

Protect your table

Heat will eventually pass through a solder board and burn an unprotected table. Work on a flameproof table or protect the table you have with something fireproof. Not everyone has a steel table. Most benches and tables are made of wood. Put something under your board, like a 12" (30.5cm) square ceramic floor tile. Keep the solder area clear of anything flammable, such as paper or plastic.

The next level of protection is to cover the tabletop with concrete tile backer board, available at hardware stores. Score it with a utility knife and then snap it to size. Want more protection? Set up a torch station with some landscaping bricks. This is a good choice for tank torches, especially if they have large tips with big, hot flames. A station like this can handle soldering, annealing, or even casting ingots. Put a sheet of steel down as a base to catch small parts. The brick wall around the sides helps keep the flame inside the work area.

Your studio setup can start with just a bench or sturdy table, with just a little organization. Protect your table with a ceramic floor tile under your solder board.

A torch station built with landscaping bricks is a good solution for use with tank torches and high heat.

Keep the flame where it belongs

The first safety rule to learn is to keep the flame where it belongs: in the soldering area. This includes igniting the torch, which some beginners think should take place while pointing the torch in the air, at the table, or onto the bench. Not so. The only surfaces that can safely take the intense heat of any torch are soldering surfaces: solder board, charcoal, and firebricks. Everything else burns: you, your table, drapes, and the carpet.

Here are examples of efficient setups: a bench (above) and a workspace (below right).

Protect your room

Let's face it: Gravity works even in the studio, so eventually something hot will fall on the floor. If you care about your floor, protect it with a remnant carpet or mat. Just confirm that anything hot that falls on your safety mat is quenched and doesn't smolder. If you set up to solder less than 3 ft. (91.4cm) from a wall, protect it with a sheet of concrete tile backer board.

Where to set up your studio

The best place to set up your studio is in an open space that has some natural ventilation, like near a window. Don't set up in a confined space, like a walk-in closet. Although this book emphasizes safer tools and chemicals, some fumes can affect your health if you solder full time for years. Solders contain zinc, and when they flow, a little is released as gas. Some flux contains fluorides, which also create fumes. Non-fluoride fluxes are a healthier alternative. Keep your face back from the solder area; hovering over the board is a one-way ticket up your nose for heat and fumes.

A simple way to increase ventilation is to work near a fan. The fan should blow air away from your solder area. Small bench-top fans with filters for fumes are also available. You can make a low-budget ventilation system with ducting from the hardware store and a window fan, or install a kitchen exhaust hood. The ducting sends any fumes out the window. Lightweight respirators and dust masks rated for minor fumes from soldering are available from jewelry suppliers.

Set up your studio near a water faucet to make it easy to clean and refill quench bowls and pickle pots. The best choice is a utility sink that is not shared with food or dishes. When my studio was in a small outbuilding in our yard, I improvised a sink from a drink cooler with a spout, and used a big bowl alongside it. Put the pickle pot near your sink or on the bench with a tray to catch spills.

Good lighting can make a big difference in your studio. Natural light is best, but it's easy to add a good lamp to your work area. Daylight matching bulbs are good for reading detail and color. One drawback to too much light is that it's difficult to see the first stages of heat on metal, as it glows a light pink, especially on silver. Dim the light or use a shade. A soldering station with bricks can have another steel sheet across the top for a hood, providing shadow while you solder or anneal.

Jewelry bench vs. a kitchen table

Wherever you work, it should help you make jewelry, not make it harder. One or two sturdy tables can make a perfectly usable work area, or you can buy a jewelry bench. Whatever you choose, it should be adaptable to the tools used for jewelry, ergonomic, and easy to organize. See my sample layout below for one arrangement idea.

It's fairly easy to improvise a bench. One of my benches was made with a tabletop set on two drawer units. I recommend solid-wood tops to withstand hammering. Your bench should be heavy or braced against a wall to keep it from walking away when you push against it, such as when you load a saw frame.

Your work area will have to adapt to making jewelry, but the needs are simple: a pin for filing and sawing, a hanger for your flex shaft, a place to solder, and a way to catch precious metal filings. Bench pins can be clamped to the table or screwed in place, usually in the center of the bench for easy access. Hang a flex shaft from a hanger above your dominant hand. Tuck the foot pedal under the table, within easy reach.

Sample Layout

• Sink   • Pickle   • Lower level for hammering

• Bench

• Bench apron for organizing tools

Soldering can share the same work area as filing and other tasks, but it's also nice to have a separate place for it so the pickle and heat are separate from your good tools.

Filing and sawing make a lot of dust, and sterling and gold dust are worth money. If you file over a tray, you may be stunned by the amount of dust that piles up! This dust, also called sweeps, can be recycled for credit or cash at a metal refinery. Jeweler's benches come with a sweeps drawer for catching metal and, with a little imagination, you can add one to a tabletop station. One solution is to wear your tray: Wear an apron that you hook to the underside of the table. The apron makes a bag underneath the pin to catch filings. Just don't forget to take it off before you get up!

Where you work can affect your body over time. Normal tables and desks can be too low for jewelry work. Sawing, filing, and even soldering is best done when the work surface is raised to counter height (about the level of your clavicles). If the work is too low, then your neck and back will hunch, causing injury and pain over time. When you are seated, your eyes should be close to the work, which promotes better posture.

Use an adjustable chair to raise and lower your body for the job at hand: Move down to be closer to the pin to see better while filing, sawing, or setting. Raise yourself a little above the board for soldering. If you're working at a table and the chair can't be lowered enough, raise just the work area: Raise the soldering area with some fire bricks, for example, and clamp the pin in a bench vise, adding a little height.

A bona fide jeweler's bench does have a lot of advantages. It already has a lot of the necessary tools such as the pin, hanger, and sweeps drawer built in or available as easy-to-install accessories. It's built to be ergonomic and it helps you organize your tools with shelves and drawers. It's like a customized toolbox with a place to work! Benches can cost under $200 or over $2000. If you're handy, you can make your own, and the leg height can be customized for you. Bench level is a little high for some tasks, like hammering, but you can stand, put the anvil in your sweeps drawer, or use a lower table.

Make your bench area work for you by making places to organize your tools within easy reach. Bench aprons can be attached that telescope out with hanging storage for files, pliers, and hammers. The shelf inside the bench can hold lengths of inexpensive plastic pipe to organize your files. And there are lots of bench accessories that save space, transforming themselves from bench pin to solder board, for example. Store more tools, metal, and more in drawers, whether they're built in or separate, and try to organize them by task. For example, put all of your burs and polishing bits in one drawer. Finally, label everything so that you know where it belongs. If a tool has a home, then it has a place to go, and your bench has a chance to be a clean, uncluttered, and efficient space.

Most jewelry benches come with a sweeps tray for filings, drawers for tools, and a bench pin.

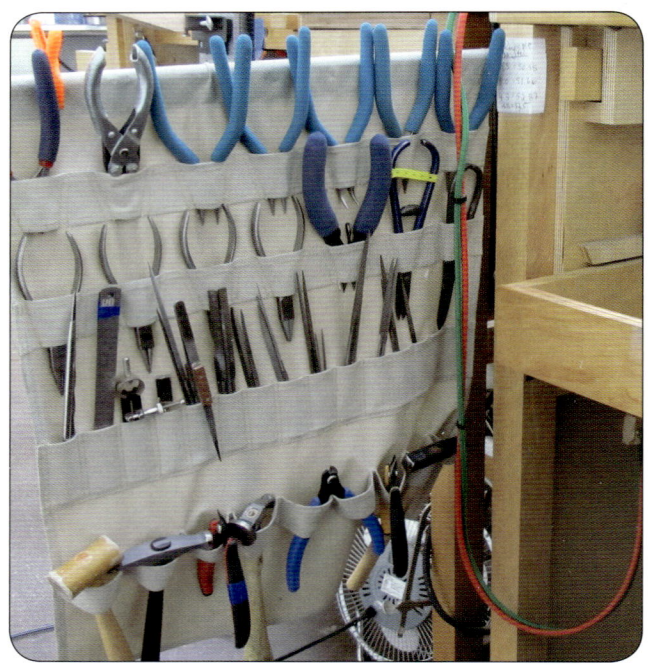

Attach a telescoping bench apron for hanging storage of files, pliers, and hammers.

Use lengths of inexpensive PVC pipe to organize metal files.

Setting up a small-torch system

Although a lot of successful soldering can be done with butane torches, at some point you may want to upgrade to a faster, hotter small torch that runs on gas tanks, like acetylene or propane. Setting up a system is simple, but it's important to do it right to avoid leaks. Read and follow the instructions that come with your torch and refer to *How to Solder*, p. 19, for important safety instructions.

Tank torches can be single gas or mixed gases. A single-gas torch that is good for jewelry is an acetylene torch. The torch draws in air to make a large flame for annealing or soldering big jobs, but it's overkill for small pieces. A mixed-gas torch uses oxygen to accelerate the fuel to higher temperatures. Acetylene is dirty before oxygen is added, making a rain of soot particles. Propane is much cleaner and recommended for a home studio. The tanks don't have to be big for jewelry—less than 2 feet tall.

Two common brands of small torch kits are the Gentec Small Torch and the Smith Little Torch. These torches have comfortable handles and interchangeable tips. The tips allow you to go from a flame smaller than a pencil point up to a large melting flame. (The companion DVD to this book demonstrates how to set up a typical small-torch system.)

A small torch can run on disposable tanks, small refillable tanks, or a hybrid of both. I like the hybrid system that uses a disposable propane tank and a refillable oxygen tank. Oxygen is used faster than fuel, so a refillable oxygen tank will last longer and is less wasteful.

Tanks are available and refillable at welding suppliers. Staffers there can be excellent resources to help you set up and maintain your tanks and torches.

Interchangeable small-torch tips range from a multijet tip (top) that creates a large melting flame to tips that make flames smaller than a pencil point.

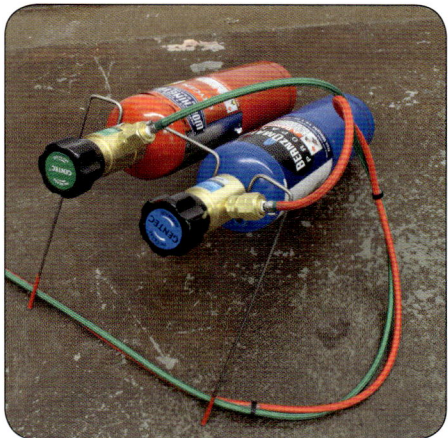 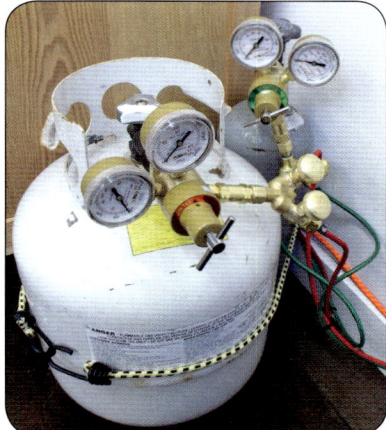 

A small-torch system can run on disposable tanks (left), small refillable tanks (center), or a hybrid of both (right).

Anatomy of a torch

Whether you are setting up a single-tank acetylene torch or a mixed-gas, oxygen/propane small torch, they share some common features. Green or blue hoses are for oxygen, and red is for fuel (propane or acetylene). The nuts and threads for attachment are specific to either fuel or oxygen lines, and they tighten in different directions so you can't attach them to the wrong tank. The oxygen line tightens clockwise, and the fuel line connections tighten counterclockwise. Fuel line connectors are notched. Similarly, a fuel regulator will often have a red color indicator and notches. There is one exception to the color system: disposable tanks. They come in colors that don't match: blue for propane and red for oxygen. But the hoses still connect red to fuel and green to oxygen, no matter what color the tank happens to be.

Assembling the torch

Make sure the tanks' valves are off before assembling or disassembling a torch. Obviously, work away from open flames or anything that can spark. The tanks should be secured with chains to something sturdy, like your bench or a wall, so that they can't be knocked over, which can damage the valves and cause an accident. Start by connecting all the parts, and then tighten them afterwards. The threads on torches are sensitive. Don't over tighten or you might damage them.

First, connect the regulators to their tanks. Tank regulators have one or two gauges, to indicate either the remaining amount of gas in the tank or the pressure in the line to the torch. Disposable tanks use their own regulators that don't have gauges. The knob opens the tank or closes it.

I recommend attaching flashback suppressors to both regulators. These are safety devices that will shut off the torch if a flame tries to go back up the hose. If they activate, the torch won't work until they are removed. In over twenty years of operating torches, I've never had one activate, but I'm glad they're there.

Next, connect the hoses to their respective regulator or suppressor. Tighten all the connections with a crescent wrench, but don't over-tighten. The torch tips are generally not attached with a wrench, which can damage them. The threads are precise enough to make a tight fit with your fingers without muscling it. Only use tips that are compatible with the type of torch and fuel. If you see flame flash around the bottom of the tip during use, turn it off and tighten it. Don't over tighten the knobs on the handle either. These needle valves are very sensitive. Just close them "finger tight."

Check for leaks

Turn on the tanks with the knobs on top. Spray soapy water or leak detection solution around the connections between the tank, regulators and torch. A leak will make bubbles. Another way to check for leaks is to turn off the tanks, but don't drain the lines. Wait and watch the regulator dials. If there's a leak, then the pressures will fall. Check the connections and then retest. Never check for gas leaks with a flame! Propane has a noticeable smell. An inexpensive safety device for your

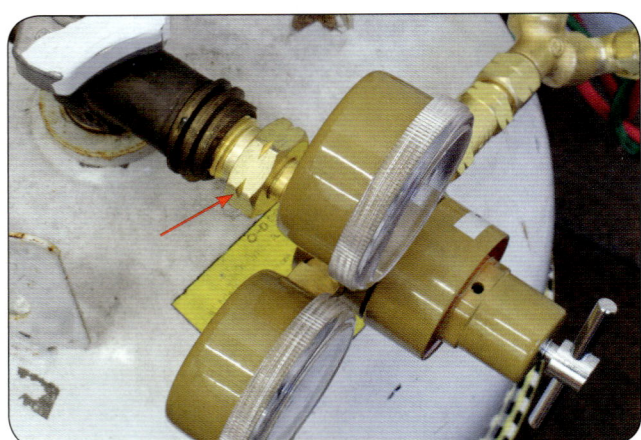

You can identify the fuel line connector by its notches.

studio is a plug-in gas detector. These will warn you if there's a leak. Check for leaks whenever you refill the tanks.

Adjust the pressure

Different size tips require different pressure settings for fuel and oxygen, and the pressure can be adjusted with the regulator. Disposable tank regulators use the pressure from the tank and are opened all the way for use. Consult the instructions that come with your torch for recommended settings. Increase the pressure by turning the T-bar in front clockwise, and decrease it counterclockwise. You'll have to open the line at the torch to release pressure for the regulator to drop. Never set the pressure higher than the recommendations. The sample pressures below are for a Gentec small torch.

Turn off the tanks and drain the lines

At the end of the day, when you're done with the torch, turn off the tanks and drain the lines for safety. Close the valves on top of each tank. Then open the fuel and oxygen lines on the torch handle. Watch the dials on the regulators as the pressure falls completely to zero. Turn off the fuel and oxygen lines on the torch handle.

TIP SIZE	OXYGEN (PSI)	FUEL (PSI)
3	6	3
4	7	3
5	8	3.5
6	9	4
7	12	4
Melting Tip	14	10

How to Solder

Soldering is used to join metal to make jewelry, and so solder is made to flow at a lower temperature than the melting point of the metals being joined. In other words, solder flows first, before your jewelry melts.

In jewelry making, we use high-temperature solder called hard solder that is made from precious metals such as sterling silver and gold. Hard solder flows at over 1200°F (649°C). This is not hardware-store or stained-glass solder, or soft solder. Soft solder melts below 800°F (427°C) and will ruin jewelry metals, making weak bonds that won't hold up to forming or annealing.

Hard solder forms very strong joins, and a little bit goes a long way. A soldered ring can be stretched up to two sizes without breaking. A tiny 1mm piece of solder wire can flow 1/2" (12mm) through a seam.

Most materials besides metal can't take the intense heat of soldering. Stones, shells, glass, and bone will crack and burn. But with the right setup, as shown in this book, successful soldering near stones is possible.

There are six simple steps to soldering, and mastering them is key to learning the art of soldering. Every time you solder, you'll repeat the same steps. With practice they become second nature. If any of the steps are skipped or not working, then that's the reason why the metal didn't solder. I've included a few FAQs (questions that my students frequently ask) to help you along.

Safety first

- Read and follow the instructions that come with your tools.
- Keep tools, chemicals, and materials away from children, untrained adults, and pets.
- Wear safety glasses.
- If anything goes wrong, turn off your torch.
- Don't release the torch or set it down while it's lit. Turn it off.
- Point your flame only at your solder board during ignition or use.
- Store butane canisters away from open flames—like your torch! Butane should not be stored in places that can exceed 100°F (38°C).
- Assume everything is hot before you pick it up, including metal, charcoal, and the torch nozzle.
- Keep water and a fire extinguisher nearby.
- Tie back loose jewelry, hair, or clothing.
- Wear natural-fiber clothing and don't use flammable cosmetics, like hairspray. Wear an apron or keep a towel on your lap.
- Wash your hands before eating and don't mix your work area or tools with food or utensils.
- Allow soldering tools to cool completely before storing.
- Make a hole to release pressure in a hollow form before soldering it closed. Without a hole, it could explode.
- Ultraviolet and infrared light from torches and hot metal can damage eyes over time or cause headaches. Butane flames are less intense, but if you're using an acetylene or propane torch, consider using shade #2 tinted safety glasses or clip-ons.

Six simple steps to soldering

1 Clean

Dirty metal can keep flux from working and solder from flowing. Solder flows best into clean, freshly sawn, filed, or sanded joins. Oils from hammering, your skin, polishing residue, and even heavy tarnish can resist soldering. Oils resist the water-based fluxes, leaving gaps that make tough firescale or prevent soldering.

The metal is clean when water flows off in sheets and doesn't puddle in drops on the metal.

Firescale, the dark scale that forms on unfluxed metals, can stop solder cold. Small pieces, like jump rings, don't have to be scrubbed unless they're in very bad shape.

Clean anything larger than a jump ring before soldering with one or more of the following methods: Soak it in pickle for 10–15 minutes, scrub it with soap and water, or rub it with a paste of water and baking soda. Rinse off any residue from cleaning, or that can interfere too.

Do solder tools and boards have to be clean too? Yes. This step also includes your tools. Baked-on flux on picks, tweezers, and soldering surfaces can make it hard to see the join or stop the solder (see *Cleaning your soldering tools*, p. 11).

2 Join

The join must be closed for solder to flow. Solder is reluctant to fill gaps, and if it does, the join is weak. But if the join is closed, with the ends so tight together that no

light shows through, then a cool thing happens: Solder is drawn through the join, from one side to the other. This vacuum at the join happens because metal gets microscopically bigger as it's heated, creating more space between the atoms. No join, no vacuum, and the solder will flow to one side or the other, but not into the join.

Joins can open during soldering if the metal anneals and it relaxes. If this happens, stop and fix the join before continuing. The projects that follow are full of helpful tips for all kinds of joins.

3 Flux

Flux keeps metal clean. Without it, metal will scale and solder won't flow. Most flux is borax-based, which forms a clear, glass-like glaze around 1100°F (593°C). This glaze protects the metal from oxygen, which, when combined with heat, forms oxides called firescale, especially in metals that contain copper in their alloy, like copper, brass, sterling silver, and some alloys of gold.

Firescale is concentrated copper, so it looks like black, gray, or red scaly blotches. Some firescale forms on the surface and can be removed by soaking in a pickle solution. But firescale can form under the surface, where the pickle can't affect it. It's harder than the rest of the metal and takes a lot of sanding to remove.

Flux saves time by stopping or minimizing firescale. Any part that isn't fluxed will get firescale, so cover the entire piece: inside, outside, the join, back, front—the works. When in doubt, flux it!

After soldering, my piece is stuck to the board. Is it the flux? Yes. Flux is like glass, so when it cools, it hardens, making it stick to charcoal and tools. Prying it off will remove hunks of charcoal or distort the annealed, soft metal. The best way to move it is to either quench it or heat it.

Small pieces, like jump rings, can be quenched with wet tweezers, cracking the flux. For larger pieces, heat them with a torch, testing with tweezers until the flux liquifies and loosens.

4 Solder

You need the right amount of solder to fill the join with minimal cleanup. Too much solder will spill out and make lumps or flow into texture or onto contrasting metals, like silver solder onto copper. It takes a little practice to learn how much solder to apply, but a good start is to always use less. Instead of cutting a big piece of solder, use a 1mm chip. If that doesn't fill the join, add more until it does. It's easier to add more than to remove too much!

Solder needs to touch the join. If it doesn't, capillary action won't draw it in. Without the vacuum, solder will flow to where the metal is hottest or where it has more contact. So if the solder is placed to the side of the join, it's more likely to flow away to that side.

5 Heat

The metal is heated to soldering temperature and the heat is balanced on both sides of the join. If there isn't enough heat from the torch or the heat is drawn away, then solder won't flow. It's the hot metal that draws in the solder, so heating just the solder won't work (see *Balancing the heat to draw solder*, p. 29).

Choosing the right surface to work on is just as important as having the right torch. Working on the wrong surface can sink the heat, drawing it out of the metal. Charcoal, honeycomb, and firebricks amplify heat, helping solder to flow faster. Solder boards take longer to heat and slow down soldering.

My solder keeps flowing away from the join, but I know it's closed. Why? If the torch is heating more from one direction, the solder will flow in that direction, even if it's away from the join. Solder is attracted to whatever part of the metal gets to solder temperature first. Adjust the direction of your flame to heat more evenly on the join line, balance heat between thick and thin pieces, and look for signs of correct temperature on both sides of the join, such as flux consistency and the color of the metal.

6 Pickle

After soldering, metal needs to be cleaned to remove flux and firescale before any more work is done, including hammering, forming, and more soldering. Old flux and firescale can dirty the metal and prevent solder from flowing.

Pickle is a cleaning solution. It can be made from dilute acids, such as sodium bisulfate or Sparex #2, or natural, biodegradable materials, like citric acid. I recommend citric acid, because it is much more

home friendly and doesn't require neutralizing. Pickle works faster when it's steaming hot, so it's heated in a pickle pot (usually an electric crock). After pickling, always rinse the metal in water and dry it before using steel tools. Copper tongs are recommended for reaching into pickle. After it's been used and starts to turn blue, a pickle like Sparex will flash copper onto metal, turning silver pink, if steel tweezers are used. Citric acid doesn't have this reaction.

A simple sieve, made by drilling small holes in the bottom of a plastic cup, is a handy way to retrieve small pieces.

Do I have to pickle after every solder join? No, just when the metal looks too dirty to solder. After a good solder join, when the flux is intact, the metal will look clean under a clear glaze of flux. After a lot of heating, the flux will start to turn green and blue, and the metal may blacken. The color is caused by copper scale, and the flux is getting saturated with it and breaking down. Blackened metal is a sign of firescale already forming. If the join looks like it's dirty or scaled, it's time to pickle.

Are there any alternatives to using pickle? Yes, but pickle is the most effective way to clean metal after soldering. In a pinch, boiling hot water will remove flux, but not firescale. Flux doesn't disappear on its own. It just gets sticky and gross.

USING SOLDER

Solder purchased from jewelry suppliers comes in different types for use with different metals, and in different forms, like wire, sheet, and paste. The most common are silver and gold solder. Silver solder is simple: With varying proportions of silver in its alloy, it is made to match the color of sterling silver and melts at different temperatures or flow points (referred to as easy, medium, and hard). It has to melt at a lower temperature than the melting point of the metal, so low-temperature zinc is alloyed with silver to make jewelry solder. The more zinc that is added, the lower the flow point and the more yellow the solder looks against silver. Hard solder is the best color match to sterling silver. Easy solder, which flows at a lower temperature, has a yellow cast. (Do not buy "silver" solder at a hardware store; purchase solder designed for jewelry making.)

Use silver solder for sterling silver, fine silver, copper, brass, bronze, nickel, silver-filled, and gold-filled. The chart below tells you silver content of various silver solder types, flow temperature, and color of heat for flow.

SILVER SOLDER	SILVER CONTENT	FLOW POINT	COLOR OF HEAT WHEN SOLDER WILL FLOW
Extra easy	56%	1207°F/653°C	Light pink
Easy	65%	1325°F/718°C	Light pink
Medium	70%	1360°F/738°C	Light red
Hard	75%	1450°F/788°C	Red
IT	80%	1490°F/810°C	Bright red

Soldering mixed metals

Silver solder will be obvious on copper or brass. Even if it stays just in the join, the solder is a different color, but a good solder seam—even if it's silver—isn't very distracting on most designs.

Gold solder is available in yellow, white, rose, and green for 8k, 10k, 14k, 18k, and 22k alloys. Every combination of color and karat gold solder is available in easy, medium, or hard. That's a lot of information to track, so gold solder is often sold as sheet, with details labeled as a code. The code stands for KARAT/COLOR/FLOW POINT. For example, 14KYE means 14k yellow easy solder. 18KGH stands for 18k green hard. 10k and 14k solder are usually a good match for gold-filled.

Other metals, like platinum and Argentium sterling, have their own special solders. Argentium solder is available in all three flow points: easy, medium, and hard. Platinum solder melts at over 1800°F (982°C) because the melting point of platinum is much higher than other jewelry metals.

Brass and copper solders can be a better match for their respective metals, but they don't have the flexibility and precision of silver solder. The solders that are available come in one flow point, which is good for making a single join in a jump ring or something similar, but can be frustrating for more complicated projects, like hollow forms, stone settings, or cuff bracelets. Copper solder melts at around 1310°F (710°C) and can still be gray in color. Heating it again after soldering and pickling

can bring copper up to the surface for a better match. Brass and bronze solder is similarly limited, with only single flow points available to date. They melt at 1200°F (649°C) for wire or 1100–1500°F (593–816°C) as paste solder. Copper and brass solders that have melting points around the temperature of tin solder cannot be mixed with your regular tools and could corrupt other metals, since they melt at such low temperatures.

A yellow alloy of silver solder is an inexpensive alternative to real gold solder for working with gold-filled. However, it has only one flow point (1500°F/816°C), which means previous joins can droop, move, or open during soldering. It can look a little silvery against the gold. I recommend 14k or 10k yellow solder for the best match.

Forms of solder

Solder comes in wire, sheet, and paste. The choice of which to use is influenced by the type of join, but it also often comes down to preference and results. The projects in this book will recommend one form over another when it makes a difference. In the case of wire or sheet, the solder is cut into tiny pieces and often melted into a ball before applying it, so it doesn't make much difference which one you start with. Solder wire is usually 20-gauge and sheet solder is 30-gauge.

Cut sheet solder into tiny pieces and melt them into balls to apply.

Paste solder comes in jars or syringes. It is a mixture of powdered solder, flux, and chemical binders. It's applied as a small blob at the join. When heated, the binder burns off, coating the join with flux. The powdered solder melts into a ball and then flows into the join. Paste solder is easy to apply but hard to control. It will often ball up away from the join, and it's easy to apply too much. But for certain jobs it's great, such as when you're soldering the end of chain into an end cap: Stuff a little paste solder inside the cap, push in the chain, and solder. Paste solder expires, drying out over time or if left uncovered. And the chemicals in it cause fumes, so use good ventilation or wear a respirator rated for solder fumes.

Marking solder

Sheet and paste solders are usually labeled with their metal and flow point. If sheet solder isn't labeled, stamp it yourself. (Solder is brittle, so stamp lightly or use a permanent marker.) Wire solder looks like regular metal, so mistakes can happen. It's easy to make jump rings out of unmarked wire solder and then watch them melt as the solder flows. Or finding that solder won't flow, even as the chip is sitting in a blaze of orange fire. Why? Because it's not solder, it's regular silver wire. Mark the end of your solder wire with a simple code of right angle bends: one bend for easy, two for medium, and three for hard.

ALL ABOUT FLUX

Flux and solder go together like ham and eggs. You can't solder without flux. Flux keeps metal clean and helps solder flow. It forms a glaze that protects metal from firescale, which can stop solder from flowing. Metals that contain copper, including copper, brass, nickel, sterling silver, and some alloys of gold, especially need protection with flux. Metals that resist scale, like fine silver and Argentium, require only a minimum of flux at the join. Flux can be grouped into three categories: flux, fire coats, and combinations of both.

Jewelry fluxes are rated for hard soldering or brazing, and they are not the same fluxes used with low-temp solders. That means that they turn clear at 1100°F (593°C) and remain effective up to 1600–1700°F (871–927°C). As flux absorbs firescale, it tends to change color, from clear to green to blue. Blue means that the flux has absorbed a lot of copper scale and it's starting to break down, a sign of overheating.

Flux

In its most basic form, flux minimizes scale and helps solder flow. Fluxes in this category include paste and self-pickling.

Paste fluxes are mixed with water to a creamy consistency and are easy to apply with a brush. Common paste flux brands include Handy, Otto, and Grifflux. Paste flux is water and borax, and can include other chemicals, such as fluorides. Fluorides can be inhaled during soldering and are harmful over long-term exposure (fluoride-free fluxes are available). Paste flux won't stop firescale, but it will minimize it if you coat the entire piece.

Self-pickling flux is liquid and can be applied with a brush or from a squeezable bottle with a needle spout. Self-pickling flux is applied with heat. After dripping it on, heat with the torch to dry it. Repeat until it has an even coat of flux. You have to be quick with the flame to catch the liquid flux before it just flows away! Self-pickling flux works best on just the join.

Fire coat

A fire coat is a firescale retardant. It acts as a strong barrier to stop scale, but it doesn't have the qualities of a flux that would be used on the join for the soldering itself. Fire coats are applied first, and then flux is applied to the join. An easy fire coat is made of a 50/50 mix of boric acid and denatured alcohol. Store the solution in a small glass jar with a tight lid to prevent evaporation. Dip the clean metal and stir it around for an even coat. Place it on the solder surface and ignite it with the torch. It will burn with a green flame and leave a flat coat of borax. If the jar catches on fire, put the lid on to snuff out the flame. Store it at a safe distance away from the torch.

Other fire coats include Stop-Ox and Prips flux, both of which are sprayed onto warm metal (see *Using spray flux*, right). Available in a ready-to-use mix, these two formulas tend to be more effective than boric acid and alcohol at stopping firescale and are nonflammable.

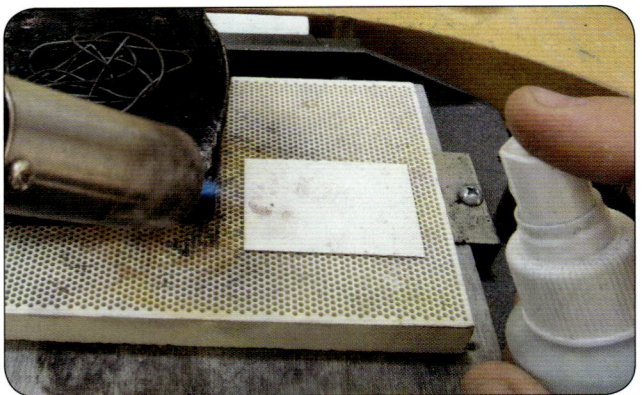

Apply spray flux as you heat with the torch.

Flux and fire coat

What could be better than something that both stops firescale and works as a flux? Well, not much when it comes to soldering! Some of my favorite flux hybrids include Cupronil, Firescoff, and Magic Boric Soldering Dip. Cupronil and Firescoff are spray-on fluxes. Both are great at stopping firescale, but Firescoff costs about four times as much as your average flux. It does a good job of stopping scale, and can be removed with just warm water after soldering, but spraying uses a lot of flux, and at that price it's a bit hard to swallow. Cupronil is almost as effective and costs a lot less. Magic Boric Soldering Dip is a nonalcohol-based flux that doesn't have to be ignited or sprayed: Just dip and solder.

Using spray flux

Spray flux has to be applied with the torch. The heat evaporates the water at 212°F (100°C), so it takes only a little heat to do it. Spray it on and heat the metal until it has a dry, flat, white coat of flux. The flux doesn't need to be as thick or lumpy as paste flux. Spritz any missing areas and dry with the torch. When one side is complete, use tweezers to turn it and coat the rest. If you see the flux turn clear, or if colors like yellow, red, cobalt, or black appear on the metal, it's too hot and you haven't used enough flux.

Spraying flux wets the solder board and surrounding area, and they may get sticky; excess flux can pool on solder boards and charcoal, making it hard to get the flux to dry on the metal. It's best to use spray flux over porous blocks, like honeycomb or firebrick. A couple of firebricks, propped up around the solder board, make good shields to contain overspray. Dry any extra flux with a towel immediately after soldering and cooling.

Spray bottles clog easily, usually during application, so clean them before the flux dries in the nozzle. Here's an excellent trick for a clog-free sprayer: turn the bottle upside down so that the tube is above the liquid. Spray repeatedly into a lined trash can until the nozzle is empty. Clean the nozzle and store it upside down.

HOW TO IGNITE YOUR TORCH

Butane torches

Most butane torches come with instructions, and they follow this simple pattern to ignite:

1. Release any safety switches [A].

2. Adjust the gas level [B]. Turn micro butane torches to maximum and large-flame torches halfway. For Blazers, turning on the gas starts the flow of butane for ignition. Other torches don't start until the trigger is pressed.

3. Press the trigger to ignite the flame [C]. Press all the way to make a spark. If it doesn't light, release the trigger and try again. Safety switches may have to be released again.

4. Engage any locking buttons to keep the flame on **[D]**.

5. Turn off the torch by pressing the trigger or releasing the lock.

Tank torches

Tank torches don't have automatic ignitors, like butane torches. Instead, a torch lighter (striker) or electric torch ignitor is used, and the torch is turned off by shutting off the knobs on the handle.

For a single-gas torch, turn on the gas and strike the flint in the cup in front of the nozzle to make a spark to ignite the flame. It takes only a little bit to light the torch, so open it just enough to hear the hiss of gas. If it takes you awhile to light it, gas can build up. Stop and wait a couple of minutes for gas to dissipate before trying again.

Mixed-gas torches, like oxy/propane, require the gases to be turned on and off in order. This same order would work for acetylene, too. Adjust the pressure at the regulators first (see *Setting Up a Studio: Setting up a small-torch system*, p. 17).

To light and turn off the torch, just remember the acronym **POOP: ON, Propane Oxygen; OFF, Oxygen Propane**.

1. Turn on the propane, often just barely cracking the valve open. Ignite with a sparker or electric ignitor. The flame should not be disconnected from the tip of the nozzle. If it is, decrease the fuel until it touches and stabilizes.

2. Slowly turn on the oxygen to modify the flame to neutral or oxidizing.

3. To turn off the torch, reverse the order: Turn off the oxygen, then turn off the propane. POOP!

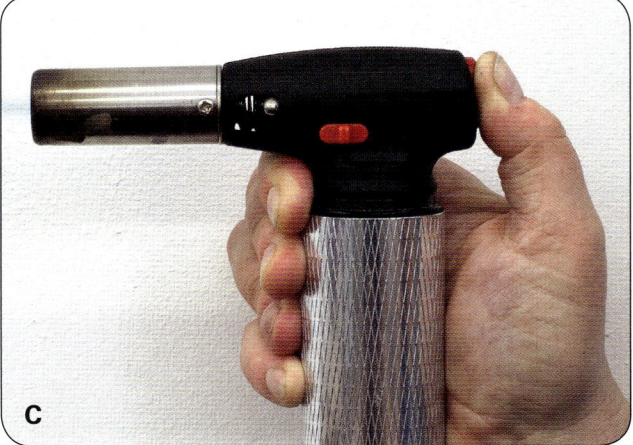

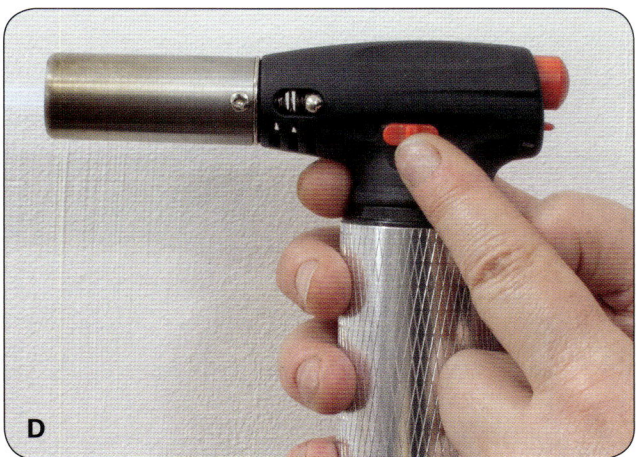

ADJUSTING THE FLAME

Using the correct size flame can help you avoid overheating the metal and causing firescale. Butane torches and tank torches can produce similar types of flames, but their level of heat is very different. Butane is colder, peaking at 2500°F (1371°C). Single- and mixed-gas torches range from 2500°F (1371°C) to 6000° (3316°C), depending on fuel types and the size of the tip. Smaller tips make hotter flames. Read the instructions that came with your torch for specifics on how to adjust your flame size. These instructions work for most torches.

Types of flames

When the gas mixes with oxygen, you'll see the outline of a cone inside the larger body of the flame. This is the air flow inside the flame, and adjusting the flame will change the appearance of the cone. A poorly adjusted flame can be all that stands between you and successful soldering. These flame types are shown with a butane torch.

Oxidizing flame
The sound of the flame is loud and the cone has a sharp point. This flame is the hottest, and it can create more firescale. This is a good flame for melting and precise, fast heating.

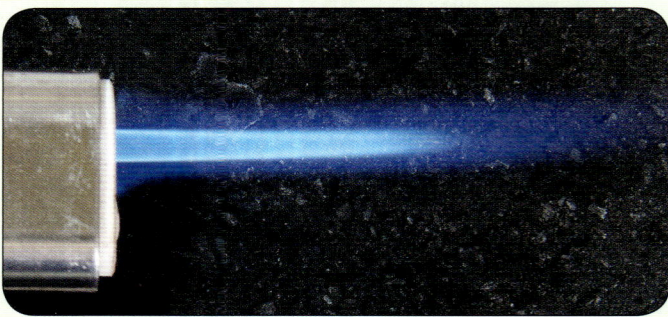

Neutral flame
A neutral flame has an even mix of oxygen and gas for a hot flame and less firescale. It is quieter than the oxidizing flame, with the outline of a cone with a feathered tip. This is a good flame for most soldering.

Reducing flame
The reducing flame is also called an annealing or bushy flame. It is rich in gas and will minimize firescale by reducing oxygen as it heats the metal. The cone is faint or very feathered, but the flame is still blue with no yellow tint. With this flame and a charcoal block, which also reduces oxygen, you can anneal with less scale. This flame can be too cool or too big for precise soldering with a butane torch.

Where's the heat?

The most heat is in front of the tip of the cone inside the flame. Moving too far into the air flow can blow solder out of place, cool it off, or slow down soldering. Butane flames are most effective with the tip of the cone close to the metal, but mixed-gas small torches are so hot that getting close to the cone will melt your metal! Instead, work farther out in the transparent feather of the flame.

Butane

These torches can be used for soldering and for making crème brûlée, which makes them very home friendly. They combine butane and air to make a focused flame. The two basic adjustments available are the amount of gas and a way to add or remove air. The gas lever or knob increases or decreases the size of the flame. Pencil and micro butane torches solder fastest with the gas turned all the way on. As the butane charge is used up, the flame will get smaller, weaker, and more transparent. The maximum flame on a charged micro butane torch, measuring the length of the cone from the base to the tip, is around 1 1/2" (38mm). A large-flame torch has a 2" (51mm) cone at maximum charge, but the width is about twice the size of the micro butane torch flame, which allows it to heat more surface area. A medium flame with the large flame torch has a 1" (26mm) cone.

If the torch has a vent control for the air, it is along the nozzle and looks like a plastic sleeve with a hole in it. Turn it to open or close a matching hole in the nozzle, allowing or cutting off air flow. Open the hole wide for an oxidizing flame; close it for a reducing flame. At around halfway open, the flame becomes neutral. Some torches, like the large-flame model, don't have good air control, and automatically make an oxidizing flame, which is acceptable for the lower temperature of butane.

Single-tank torches

Single-tank torches draw in air like a butane torch. They include propane and acetylene torches. The flame can sometimes be adjusted to make one of the three different flame types, but usually it automatically makes an oxidizing flame that can be increased or decreased in size with the gas knob.

Mixed-gas torches

Mixed-gas torches, or small torches, use a fuel and oxygen mix to make hotter flames. The flame is modified by increasing and decreasing the oxygen. More oxygen will sharpen it to an oxidizing flame. Less oxygen will soften it to a neutral flame. It's hard to get a reducing flame with small torches before they just turn yellow, which is too dirty and full of soot to use. These instructions are for small oxy/propane or oxy/acetylene torches.

To adjust the size and type of the flame, whether you're using a small tip or a melting tip, work with the gases in the same order described in *How to Ignite Your Torch*, p. 23. The size of the flame affects how much surface can be heated effectively. Soften the flame for delicate work; increase it for faster heating of larger pieces.

To increase the size of the flame: Increase the fuel by an eighth of a turn. Then increase the oxygen to return to a neutral or oxidizing flame. To increase the size of the flame further, repeat this step.

To decrease the size of the flame: Decrease the oxygen by an eighth of a turn, and then decrease the fuel to return to a neutral or oxidizing flame. Repeat until you have the size flame you want.

For small tip (#3–7), a small flame can be just 2 or 3mm from the base of the cone to the tip. A medium flame is 4–5mm, and a large flame is 6mm or more. The sound of the flame will get louder as it gets larger and hotter.

A melting tip makes a bigger, wider flame composed of six individual flames. A small flame is about 3mm long from base to tip of the cone. A medium flame is 10mm when the cones are in neutral mode, and a large flame is anything longer than 10mm.

See the photos below for a comparison of flames using a small #5 tip and a large melting tip.

Oxidizing flame, #5 tip

Neutral flame, #5 tip

Oxidizing flame, melting tip

Neutral flame, melting tip

SOLDERING TECHNIQUES

Soldering involves following the Six Simple Steps (see p. 19), but using different techniques will help you apply the solder and balance the heat to make the solder flow. Some of these are adapted to the kind of solder: sheet or paste, for example. Or it may be a general technique, like pick-and-ball soldering, that can be used with wire or sheet solder. Each of these exercises is demonstrated on a closed jump ring that has been fluxed on all sides and then placed on a solder board. Soldering is two-handed, since you will need to move solder as you heat the metal. Hold the torch with your nondominant hand, and use your dominant hand for controlling the pick. I'm right handed, so the torch is in my left hand and the pick is in my right.

Closing jump rings for soldering

The join must be closed for soldering. If you hold it up to light, you shouldn't see any gap or light. Follow these instructions to close your jump ring with tension for a tight fit. Make sure the ends are aligned in all directions for a seamless join.

1. Hold the jump ring with two pairs of pliers (chainnose and either flatnose pliers or bentnose pliers). Don't cover too much of the jump ring with the pliers or they'll get in the way.

2. Open the jump ring sideways [A].
3. Overlap the ends of the jump ring by "walking" them inward **[B]**. When you bring the join together, the ends should overlap and bump against each other. If they pass without touching, try again.
4. Continue closing the jump ring, feeling the ends rubbing against each other **[C]**. Hold the jump ring up to the light. If you can see light through the join, repeat steps 1–4.

If the ring is cut poorly, it will never close properly. Hold the jump ring in a ring clamp with the join closed and saw through it with a #0 saw blade, or open it and fix it with a flat needle file.

Use the flux to help you

Understanding how the condition of the flux affects solder placement can make the difference between success and a brand new facial tic.

Sometimes application is as simple as placing a chip of solder on the join while the flux is wet, and heating everything slowly. But often as the water boils away, the solder will float and move out of position, over and over again.

Usually the best method is to place the solder after the flux has turned into a hot, sticky, clear glaze. In fact, painting a little flux on your solder before you heat it up makes it sticky and easier to scoop up.

Flux dries when its water or alcohol evaporates, then liquifies into a clear, enamel-like glaze. Paste flux takes longer to clear, and can mushroom into gooey puffs before settling down. When the flux turns clear and the metal looks clean, it's sticky enough to grab the solder from your pick. Chips placed in wet solder can fall off as the flux bubbles. Keep the flux warm with the torch, or it will cool and harden like glass, and solder won't stick to it. Remember, it's not about heating the solder on the pick; you're trying to heat the flux on the metal so it takes the solder from the pick as you touch it to the join.

Cupronil helped with the finishing steps of this cuff (see p. 63).

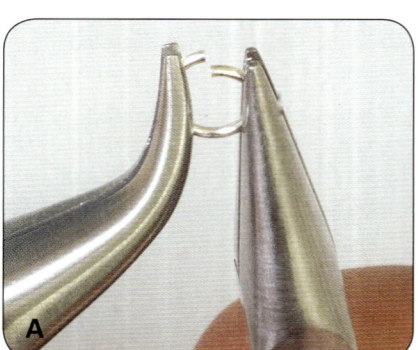

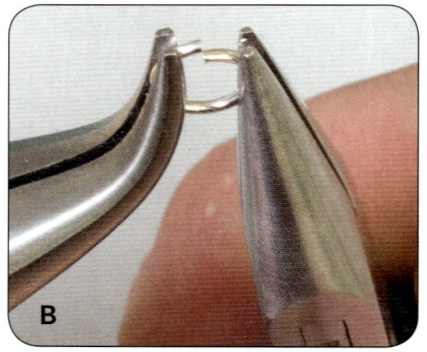

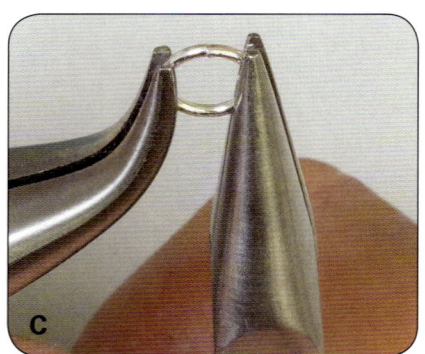

Pick-and-ball soldering

Balling up either sheet or wire solder has advantages. Flat chips can ball up during heating and move away from the join. Balls are small and easy to place right on tiny joins, like jump rings. And with a little practice, balling and scooping can be a quick way to place solder with precision.

The fluxed jump rings are placed with the joins at 2 o'clock (10 o'clock for a left-handed person). This is ergonomic, allowing the flame to heat the jump ring from the back as the solder is placed on the join.

Mark the join with permanent marker to help you see the join. Flux the entire jump ring **[A]**. Cut a 1mm chip of medium silver solder wire with flush cutters with the flat side up, to make it easy to see the measurement **[B]**. Cutting just below the bevels left on the tip of the wire from the previous cut will make a perfect size chip for most joins. Place a finger on top to catch the chip as you cut **[C]**. Place the chip on the solder board and lightly brush flux on it to keep the solder from scaling and to help the chip stick to the pick **[D]**. Heat the chip until it turns into a ball and glistens like mercury **[E]**. If it glows red, it's too hot. Keep the pick nearby, but not hot. As the chip balls up, move the flame aside and scoop it up with the tip of the pick **[F]**. If you wait too long, the flux will freeze it to the board. Too soon, and the ball will be too molten to scoop.

Warm the jump ring with the flame. When the flux clears, simmer the heat as you place the solder on the join. When the flux is sticky, it will take the ball from the pick. If the ball moves, keep the jump ring warm with the torch, but below solder temperature or red heat. This keeps the flux liquid enough while you nudge the ball into position with the pick **[G]**. Continue heating, straight through the middle of ring, in line with the join, until the solder flows.

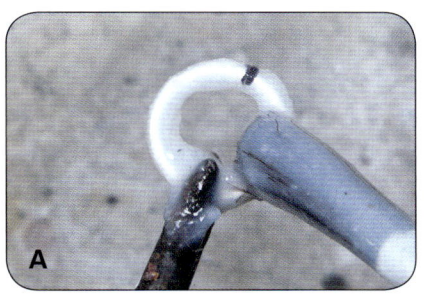

A

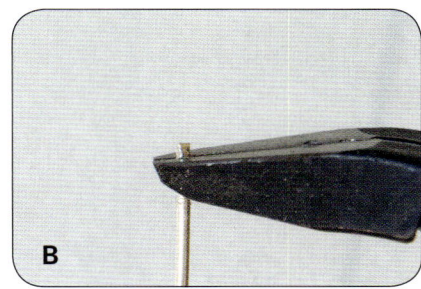

B

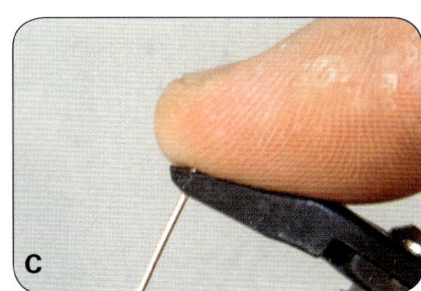

C

D

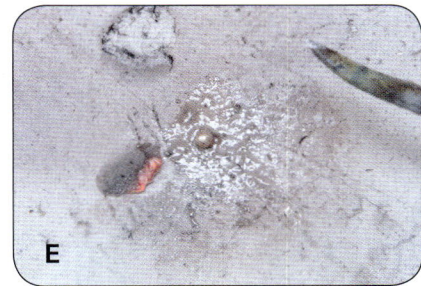

E

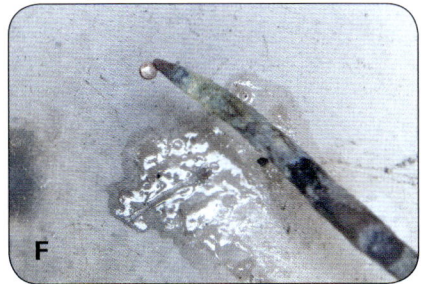

F

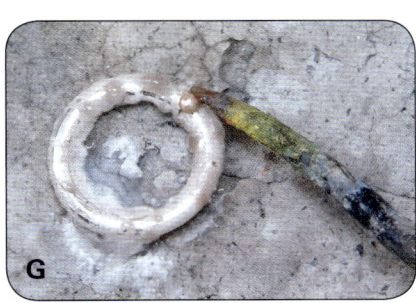

G

Chip (pallion) soldering

Use straight shears to make parallel cuts in sheet solder about 1mm apart [A]. Make a perpendicular cut across the strips to create small chips (also called pallions). Place your finger against the sheet as you cut, or cut into a small, labeled container.

Place a few chips of medium solder on the solder board and brush a little flux on each. Flux the entire jump ring [B] and lay it flat on the charcoal block, with the join facing you. Ball up the solder chips to scoop and place (see *Pick-and-ball soldering*, p. 27), or use them flat.

The chip has to touch the join. Either place it under the join before heating the flux [C], or use tweezers to place it on top when the flux is clear. Continue to heat the ring evenly, keeping the solder in place with the pick [D]

until it flows. The advantage of pinning the solder under the join is that it's a lot harder for it to move out of position.

Paste soldering

Paste solder has extra chemicals in the binder, so I recommend wearing a mask rated for solder fumes. Most products have material safety data sheets (MSDS), so you can check those for details on any health risks.

Just because paste solder contains flux doesn't mean it will protect the entire piece from firescale. Apply a flux or fire coat to the ring first, and heat it until dry or clear. Scoop a tiny bit of medium silver solder paste onto a pick and place it on the join [E]. (Even if you're using a syringe, it can be tricky not to apply too much solder.) Heat the jump ring until the paste evaporates and the solder balls up [F]. Move

the ball if it's not touching the join, or continue heating until the solder flows.

Clean the pick before using a different type of solder or use a separate pick for each flow point (easy, medium, and hard) to avoid mixing them by accident.

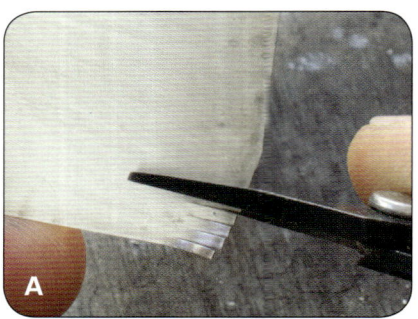

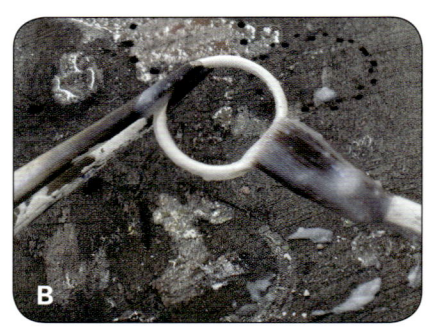

Splitting solder

A little bit of solder can flow a long way. I've seen 1mm chips flow up to ½" (12mm) along a seam when the join is good enough. Even that size can leave extra solder to clean up. It's hard to cut a chip any smaller, but chips can be split with a pick while they're molten: Cut and flux a 1mm chip of medium solder on the solder board **[A]**. Balance the tip of your pick on top, with just a little pressure. Heat it up, and when it becomes molten, press the pick down, separating it into two pieces **[B]**. Remove the heat and hold the pick in place so the halves don't slip back together.

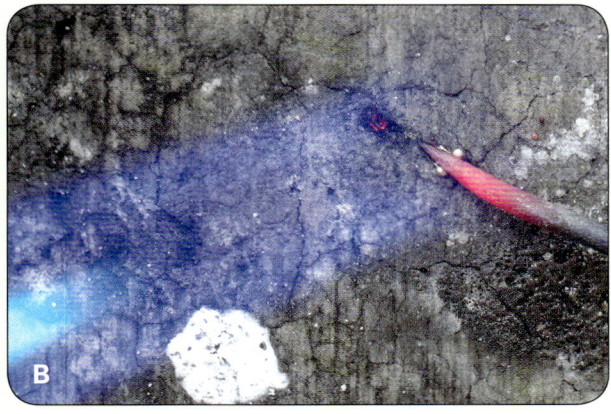

Balancing heat to draw solder

When solder fails to flow into the join, two common reasons are that it wasn't touching the join or the heat was unbalanced. If the solder is away from the join, it will flow onto the metal where it's touching. If the heat is uneven, it will flow toward the side that reaches solder temperature first. As you learn how to aim the flame, practice adjusting its direction to draw the solder in the direction you want it to flow.

Jewelry metals such as silver, gold, and copper are very good at conducting heat. Heating just the join usually fails because the rest of the metal is drawing the heat away. Instead, heat the entire piece evenly, looking for signs of heat, like the condition of the flux and any visible glow in the metal. Flux turns clear at 1100°F (593°C), just a few hundred degrees below the temperature at which solder flows. Look for the flux to clear on the entire piece before heating the join. Solder will flow between light red and bright red heat (see *Solder comparison chart*, p. 21). Look for even color on both sides of the join.

Sometimes the join is between different size parts, like soldering a small jump ring to the back of a bigger charm. Heating directly on the join will bring the ring up to temperature first, and the solder will flow onto it, but not into the join. The ring will fall off in the water or pickle. The temperature has to be balanced between the two by mostly heating the larger charm and staying away from the small ring until the very end, when the solder is about to flow.

Inspect your joins with a loupe

How do you know if the solder worked? Well, if the join doesn't open or the pieces don't fall apart, then that's a good sign. But is the join complete, or is it just tacked together? A weak join can open later. Use a 10x magnification jewelers loupe to inspect it. Hold the loupe against your eye or glasses. Bring the join up close, about an inch away from the lens, until it comes into focus **[A]**. This microscopic view will reveal any pits, gaps or a beautiful join in a way that is almost impossible to see with the naked eye alone **[inset]**.

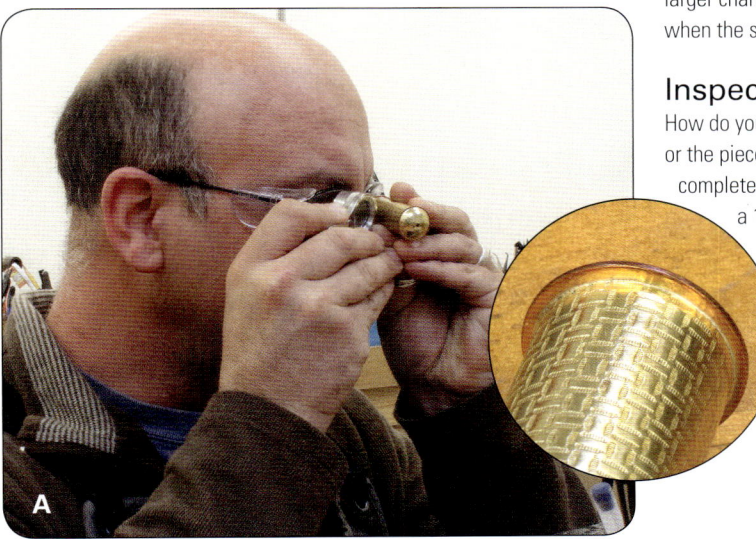

UNSOLDERING

Sooner or later, you'll solder something in the wrong place. What can be soldered can be unsoldered. Follow the same basic steps as soldering. Flux everything, as usual. When the old solder flows again, pull the pieces apart with tweezers or a pick.

It's easier if you hold one part down with a third hand or pin it to a charcoal block with binding wire. Afterward, pickle and set up for soldering again. Some solder is probably still on the pieces, so try soldering without adding any solder first.

ANTI-FLUX

Anti-flux is a material that can be painted on to dirty the metal on purpose to stop solder from flowing. It can keep solder out of texture or make it easier to clean up after soldering mixed metals. And it can slow down the flow of solder with the same flow point in previous joins as you work on a new one.

A good anti-flux is anything that will stay dirty during soldering but won't interfere with the real join. Commercial anti-fluxes are available, but substitutes are easy to find at local stores or to make. White correction fluid will work, but the oil-based version is toxic when burnt (the water-based version is safer). A paste of rouge polishing dust and water will work nicely, but since I recommend using clean, dust-free polishing attachments, you might not have any rouge on hand.

I recommend a small tube of yellow ochre gouache paint, available at art stores. It lasts a long time and the gouache burns without much odor. Apply yellow ochre with a brush, straight from the tube or with a little water. Don't mix it with your flux brush or regular water on the bench, since it can contaminate the flux.

Apply anti-flux after drying or clearing any flux or fire coat (otherwise it can mix with wet flux and accidentally spread into the join, where it will do its job and stop the solder from flowing). After soldering, let it air-cool or quench in a separate water dish. Scrub off as much yellow ochre as possible before pickling.

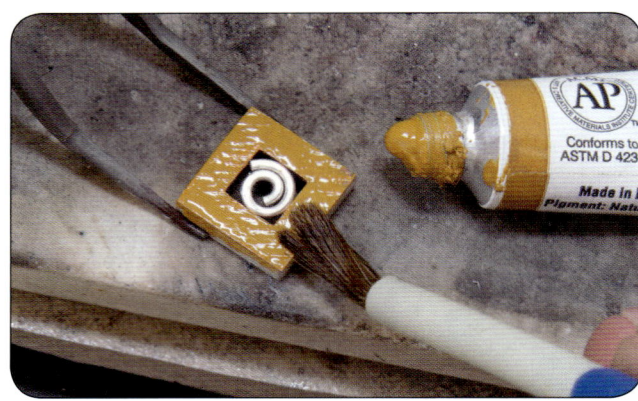

Apply yellow ochre gouache straight from the tube or with a little water.

SOLDERING NEAR STONES

Most stones can't withstand the intense heat of soldering. Even durable stones can crack if heat is applied or removed too quickly. Yet, once friends and family learn about your interest in soldering, they'll give you jewelry to repair—most of it with stones in settings. Since jewelry metals conduct heat so well, it's difficult but not impossible to shield them from the heat with these simple strategies.

Keep it away from heat

If delicate materials are far enough away from heat and only loosely connected to the metal being soldered, they can survive without any shielding (see *Beads and Rings Earrings*, p. 46). Keep the material far enough from the join to not touch the flame, and point the torch away from it at all times. The gauge of the metal being soldered can't be beyond the ability of the torch to heat quickly and effectively. And the metal should not be directly connected to the stone.

For example, a stone in a setting soldered to a ring will still burn. But a bead that can turn easily on a jump ring is OK, because the metal is loose enough to heat quickly for soldering but not tight enough to conduct the heat into the bead. Dip only the metal in the pickle, or pickle for less than

Many materials can survive without any shielding if you keep them far from heat and loosely connected to the metal being soldered.

Certain stones in settings can take the heat of the torch for a limited time.

a minute. Some of these materials are sensitive to even mild pickles. Choose durable materials, like the ones suggested in this table.

Use heat-resistant stones

Some stones can take the heat of the torch for a limited amount of time. They can be in settings during soldering. The most durable are diamonds, rubies, sapphires, and most cubic zirconia.* Too much heat applied too suddenly or for too long can damage or change the color of even the most durable stones. Heat mostly the metal, not directly on the stone, allowing it to conduct indirectly to the join. Use lower temperature solders, like extra-easy or easy. Warm everything slowly, including the stones, to avoid shocking them.

HEAT-RESISTANT STONES		
Rubies	Peridot	Alexandrite
Sapphires	Diamonds	Garnets
Cubic zirconia*	Tanzanite	Hematite

*Although most CZs can withstand soldering temperature, some cubic zirconia are sensitive to heat, including CZ emeralds and CZ tanzanites.

Never quench after soldering when stones are involved! Let the piece air-cool.

DURABILITY TO HEAT			
DURABLE	**NORMAL**	**DELICATE**	**AVOID**
Rubies	Glass	Pearls	Plastic
Sapphires	Ceramic	Pewter	Amber
Diamonds	Stone	Plated	Fabric
Cubic Zirconia	Crystal	Coated	Resin
		Opal	Wood
			Bone

Heat shields

When I worked in a repair shop we had to repair rings with set stones, including turquoise and sterling silver rings. The setting was immersed in wet sand or under water while the join in the shank was soldered. The ring would get so hot as the sterling tried to conduct heat into the setting that the steam would snuff out the flame.

Heat shields insulate delicate materials from the torch and resist the conduction of heat. These days you'll find lots of brands on the market, including Vigor Heat Shield, Kool Jewel, and Thermaguard. I prefer Vigor Heat Shield because it is completely nontoxic. Another option is to use a cool cup—a stainless dish with built-in tweezers—to submerge stones in water or wet sand.

Heat shield can be applied before or after flux (see *Fused Ball Pearl Earrings*, p. 82). Pack it completely around any stones, but leave as much of the metal around the join open and available to the torch as possible. Since a heat shield is a heat sink, too much will slow down soldering or stop it completely. Butane torches may not be able to overcome the heat sink if the gauge is too thick, but usually work fine if the metal is 1.6mm/14-gauge or thinner. The intense flame of a #3–5 tip with a small torch makes fast work of soldering the shank on a ring. Keep the flame pointed away from the heat shield, but heat all of the metal outside of it. (For more information on how to size a ring, see *Adjusting ring size*, p. 108).

A heat shield is a heat sink. A butane torch will probably work fine as long as the metal is 1.6mm/14-gauge or thinner.

SOLDERING MIXED METALS

The techniques for soldering mixed metals are the same as normal soldering, but take precautions to avoid melting metals, limiting solder overflow, and disguising join lines.

Mixing metals can lower the melting point

Metals have different melting points. Sterling silver melts at 1640°F (893°C) and copper at 1981°F (1083°C). Choose a solder that melts before the lowest-temperature metal in your piece. Another good example is soldering sterling and 14k yellow gold together: Some alloys of 14k yellow gold melt at 1476°F (802°C), just above hard silver solder, so I wouldn't use anything higher than medium silver solder or I would use 14KY solders.

But an interesting thing happens when you join mixed metals. Their melting points at the point of contact can lower dramatically, actually encouraging melting to happen faster. This is called a eutectic reaction. Brass and sterling is a classic combination. Brass can melt through sterling during soldering, as if the silver is suddenly permeable. I've seen brass walls of a hollow bead slip intact through sterling sheet and brass dots melt through silver like acid. The same thing can happen with copper and sterling. As the copper turns a deep red, sterling can flow over the surface like solder before it even glistens with any signs of melting. When soldering these metals together, don't use hard solder. Use medium or easy and avoid overheating.

Overflow

It takes a bit of work to remove solder overflow on copper or brass. It's even more delicate if silver solder gets on yellow gold! Avoiding it in the first place is better. Here are a few handy tips:

1. Use less solder. Less solder equals less mess. If you use big pieces of solder, the extra has to go somewhere.
2. Direct the solder with the torch. Solder is attracted to heat. Watch the solder as it's about to flow and turn the torch as needed to keep the solder in the join.
3. Limit overflow with anti-flux.

Disguising joins

Even the best join with silver solder on brass or copper will look like a gray line. How can you hide it? One way is to use a patina. Patina will darken the join along with the rest of the metal. When soldering silver to other metals, the silver solder will blend easily with the silver side of the join.

Soldering copper has a unique solution: Plate the join with copper to match. Sparex pickle will reverse its job and plate copper onto your metal if steel is put into the solution. Fill a separate dish with hot used Sparex. The pickle should be blue with saturated copper. Soak your metal in it with some steel wire. Remove after a few minutes when the joins have been plated. The layer is very thin, so polish your metal before plating and bring back the shine with nonabrasive finishing, like burnishing with stainless steel shot in a tumbler.

STONE-SETTING TECHNIQUES

Capturing beautiful stones in metal is the motivation for most students to learn how to solder. Many projects in this book use bezel, prong, flush, and tube settings to set gemstones, cabochons, and found objects.

Anatomy of a stone

To set a stone, you have to understand how it is cut. A cabochon is simple: It has a flat bottom with sides that taper slightly as they round toward the peak at the top. But faceted stones are cut and machined, and have labeled parts: their very own anatomical system! The illustration on the right shows the names for each section of a typical faceted stone. The closer a stone is cut to its ideal, as in this diagram, the easier it is to set.

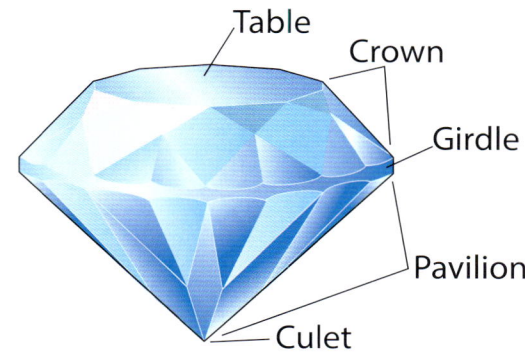

Hardness

Every stone has a rating on the Mohs Hardness Scale, from 1–10. One is the softest and easiest to scratch, and 10 the hardest. Some sample stones and their hardness are listed in the chart on the right. Durable, scratch-resistant stones are recommended for your first settings. The Mohs rating doesn't tell you whether or not the stone is easy to crack, like opal, or whether it flakes, like amber. You have to study stones and learn by experience.

MOHS HARDNESS SCALE			
1	Talc	6	Garnet
2	Pearl, amber, serpentine, ivory	7	Jasper, agate, amethyst
3	Coral	8	Chrysoberyl
4	Jet	9	Corundum, zirconia, ruby
5	Malachite, turquoise, opal	10	Diamond

Bezel-setting

A bezel is a strip of metal that forms a collar around the stone. Bezels are very strong and hard to remove; you'd have to saw off the bezel to get the stone out. They are one of the oldest styles of setting and the best one to start with as a student. Bezel wire is formed around the stone, trimmed, and soldered closed. The bezel is soldered in place. After all the fabrication and polishing are done, the bezel is crimped down around the stone and burnished to a smooth finish (see *Looking Sharp Bezels*, p. 51).

Cabochons are the easiest to set with bezels. A well-cut cabochon tapers from the edge as it rises to the top of the stone. The bottom should be flat with little or no rounding; cabochons with a round bottom are called double cut or lentils. Bullets are tall cabochons. Some cabochons are cut very thin, with narrow edges and usually need to be raised inside the setting with a step or padding.

The bezel wire that's easiest to use is made from soft metal, like fine silver or 22k yellow gold, and is around 28- or 30-gauge. It's available with straight, scalloped, or serrated edges. Stepped bezel wire, which makes its own base, or bezels with built-in borders and intricate pierced patterns, are usually made of harder metal such as sterling silver. Bezel strip is sold in different heights including ⅛" (3mm), ³⁄₁₆" (5mm), and ¼" (6.5mm).

Hold a ruler or bezel wire against the edge of your stone and close one eye as you measure it. The height of the bezel wire should be a little above

This stone is held in place with a bezel setting.

where the edge of the stone starts to taper toward the top of the stone. If your stone is too short for the only bezel you have, you can carefully shear the wire to size or raise your stone with a little padding made from plastic cut from a yogurt lid.

Prongs

Prong settings come in lots of different styles, including crown, pedestal, and basket. Imagine a prong setting as a bezel, except all of the collar is removed except for a few key places that anchor the stone in place. The advantage of prongs is that they can be pulled back with a prong lifter and the stone removed for repair work. Making a pedestal setting will teach you about how prongs are positioned and modified to set stones (see *Pedestal Prong Pendant*, p. 65). The pedestal supports the stone. If it's for a flat cabochon, the top of the pedestal is flat like the bottom of the stone. For a faceted stone, the seat is beveled on the inside at an angle to match the pavilion, so that the girdle sits flush on top of the pedestal.

A crown prong setting can be purchased, and the prongs are modified with burs to make a seat for the stone (see *Crown Prong Rings*, p. 74).

Prongs are usually in symmetrical pairs around the stone. The facets of the table of the stone form two interlocked squares. The corners of one of these squares should line up with the prongs (this is called *squaring the facets*). The height of the prongs start level with the table. When set, they are pressed against the crown of the stone, holding it firmly against the pedestal. Usually prongs look best when trimmed to end halfway up the crown or a little less.

Tube-setting

Tube settings are bezels. Some premade tube settings can be purchased in sterling or gold. But custom sizes can be made from tubing with setting burs, and are ideal for making settings for small (2mm, 3mm, and 4mm), faceted stones. Round stones are best, because they match the tubing. Whenever possible, bur the setting first, trim it to size, and solder it in place. Set the stone after polishing.

A basket prong setting has wires around the outside of the stone or forming a framework underneath it.

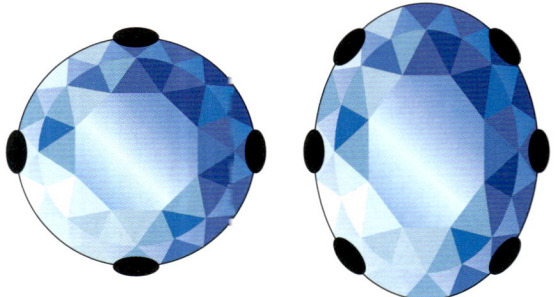
A typical prong setting has four or six prongs.

Prong height can be customized to hold found objects or irregularly shaped stones.

Purchased crown prong settings hold faceted gemstones.

Tube-set garnets in a pendant.

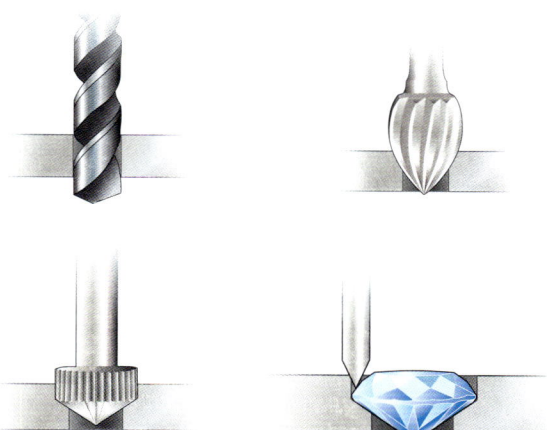

Four steps of flush-setting (clockwise from upper left): drill, use a bud bur to enlarge, use a setting bur, and burnish.

To make a tube setting, the setting bur and stone should be a close match in diameter, within .1mm. The tubing has to be a little larger to have enough material to make the bezel, and a little smaller on the inside to support the stone. Heavy wall tubing is preferred. The seat is cut with the setting bur, so that the table is slightly higher or level with the top of the vertical wall, or bezel **[figure, right]** (see *Gold-Filled Tube Set Rings*, p. 85). The bezel needs to be thin enough to easily push over without too much force, and can be adjusted with a file. The tubing is trimmed to size so that the culet or tip of the stone sits inside the setting. If it sticks out, it's sharp and can be easily chipped.

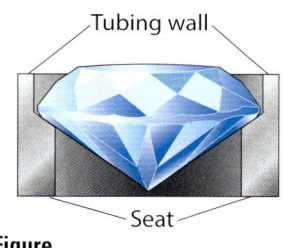

Figure

Flush-setting

A flush setting is made directly in metal, usually after all the fabrication is complete, including soldering and most of the polishing. The metal is burnished around the stone, setting it. The table of the stone sits a little above the metal or level with it. When finished, the metal looks smooth and perfect around it (see *Flush-Set Gems Cuff*, p. 93). Done right, the setting looks like a little miracle!

Flush settings take practice to master. Make a few in a small piece of 12-gauge copper sheet (or sweat-solder two pieces of 18-gauge sheet together to make enough thickness for the height of the stone from the culet to the table). Sterling and yellow gold are easier to burnish than copper. Burnishing is hard on stones, so the hardness rating should be 8 or higher. Be especially gentle with soft, colored stones.

There are lots of variations on how to make a flush setting, but these steps work well:

1. Drill a pilot hole. Center punch. Use a drill that is half the size of the stone. Lubricate the drill with bur lubricant.

2. Enlarge with a bud bur. Enlarge the pilot hole with a bud bur that is ⅔–¾ the size of the stone. Pass the bur all the way through the hole. Removing excess metal makes the setting bur cut more efficiently.

3. Cut a seat with a setting bur. Use a bur that is the same size or slightly smaller than the diameter of the girdle. Cut a clear vertical wall and pavilion shelf in the bearing. The height of the vertical wall should be half the height of the crown. Too deep and the metal will need to be tapped with a punch to set it. Too shallow and the burnisher can slip out and scratch the metal or the stone won't set. Remove any flashing from the bottom of the hole with a setting bur before setting the stone.

4. Burnish the setting. Use a needle burnisher made from an old bur to set the stone. The tip should be pointed and polished. Dull the point a little with a blue silicone polishing wheel, since it rides along the facets of the crown. Lubricate the burnisher with a light oil or bur lubricant. Start the setting by burnishing four opposite points, as you would a bezel, with the burnisher at a 45-degree angle. Raise it to be vertical, and press slightly outwards as you trace the girdle of the stone. The pressure will burnish the metal down against the crown, trapping it.

Check the setting by pressing up through the pilot hole to see if the stone moves or pops out. If a few rounds of needle burnishing don't set the stone, then use a regular steel burnisher to burnish the flat metal around the stone, moving it inward. It should visibly close around the stone. When the stone is set and won't pop out, repeat a couple of rounds with the needle burnisher to make a bright rim around the inside of the setting.

The burnisher can make a smooth but streaky shine. Use black and blue silicone wheels to carefully polish around the setting. Be careful: Some stones can be scratched by these wheels.

Right: Terms for the different angles of the bearing cut by setting burs.

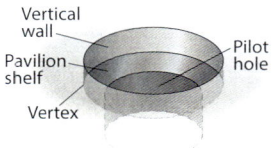

Toolkits

Use these lists as a general guide to assemble the toolkits and supplies you'll need to make the projects. You won't use every tool in every project, but this is a well-rounded list of what you'll need for jewelry making.

Safety & magnification
- Safety glasses
- Apron
- #2 shade clip-ons or safety glasses (for small torch)
- 2x–5x clip-ons or headband magnifiers
- Jeweler's loupe (10x)

General jewelry
- Jeweler's saw frame
- Saw blades: size 2, 2/0, 4/0
- Cut Lube or beeswax
- Half-round hand file with handle, cut 0, 2, and 4
- Sandpaper: wet/dry for metal, grits 180, 320, 600
- Micron-graded sandpaper: 1200-grit
- Needle files, cut 2 (set of 6 or 12)
- Steel bench block
- Nylon plastic block (or hockey puck)
- Ring mandrel
- Oval stepped bracelet mandrel
- Straight shears
- Drill bits for metal: 1/8", 3/32", 1mm, 1.6mm/1/16"
- V-slot bench pin (with clamp optional)
- Ring clamp
- Two-hole metal punch
- Bench vise
- Bezel mandrel, round
- Center punch
- Tube-cutting jig
- Scraper

Jewelry pliers
- Flush cutters (heavy-duty razor-flush recommended)
- Chainnose
- Roundnose
- Bent or flatnose
- Flat/half-round (steel)
- Nylon flatnose
- Nylon-jaw bracelet-forming pliers

Measuring
- Digital calipers
- Finger gauges
- Dividers with lock-nut
- Metric ruler
- Scribe

Hammers & stamps
- Liner stamp
- Chasing hammer
- Goldsmith's hammer, 4 oz. (polished)
- Planishing hammer, 4 oz. (polished)
- Rawhide mallet, 2" (51mm) face
- Texture hammer
- Nail sets with cupped ends: 3mm, 4–5mm, 1.2mm

Soldering
- Solder board
- Charcoal (soft)
- Honeycomb or firebrick
- Handy Flux paste flux
- Cupronil with spray bottle
- Self-pickling flux
- Flux bottle with needle spout
- Boric acid and denatured alcohol
- Titanium soldering pick
- Third hand (2)
- 19-gauge annealed-steel binding wire
- Soldering tripod
- Inexpensive small artist brush for flux
- Pickle pot (quart size electric crock)
- Citric pickle
- Copper pickle tongs
- Vigor heat shield
- Yellow ochre gouache paint (small tube)
- Ceramic ring stand

Torches
- Micro butane torch
- Large-flame butane torch
- Small torch (oxygen/propane recommended)
- Tips for small torch: #3–7, melting tip
- Manual torch lighter or electric ignitor (for small torch)

Polishing
- Flex shaft polishing motor
- Silicone polishing wheels: white, black, blue, pink
- Silicone cylinders: white, black, blue, pink
- Threaded mandrels (12)
- Mini-reinforced mandrels (12)
- Split mandrels
- Radial bristle disks, 3/4"/19mm (12 each): white, red, blue, peach, light green
- Pin polishers (2 or 3mm): gray, brown, green
- Pin holder (2 or 3mm)
- Polishing pads
- Scotchbrite pads
- Tumbler and mixed stainless steel shot

Setting
- Bud burs: 1.6mm, 2.3mm, 2.5mm
- Setting burs: 2mm, 3mm, 5mm
- Ball bur: 3mm, 1mm
- Cup bur: 2mm
- Needle burnisher
- Burnisher
- Prong pusher
- Prong lifter
- Leather scraps

2

Projects

MIXED METAL EARRINGS

Making this simple pair of earrings combines soldering jump rings and assembling them into custom findings. This project is a great introduction to the basics of soldering mixed metals: controlling solder flow and working with metals that have different melting points. Save money as you hone your skills by substituting brass jump rings for sterling silver: The techniques are the same. Judge your progress by how little you have to clean up the metal when you're done.

LEVEL
Beginner

TECHNIQUES
Soldering jump rings

Soldering mixed metals

Filing a taper

Melting wire into balls

MATERIALS
- **8** 4mm ID 18-gauge sterling jump rings
- **2** 5mm ID 18-gauge sterling jump rings
- **6"** 16-gauge copper round wire, dead soft
- Easy, medium, and hard wire solder
- **2** 10mm oval carnelian beads
- **6** 3mm round gemstone beads
- **6** ball-end headpins
- Pair of sterling silver earring wires

TORCH
Micro butane torch, maximum flame, or small torch with #5 tip, medium flame

FLUX
Paste flux

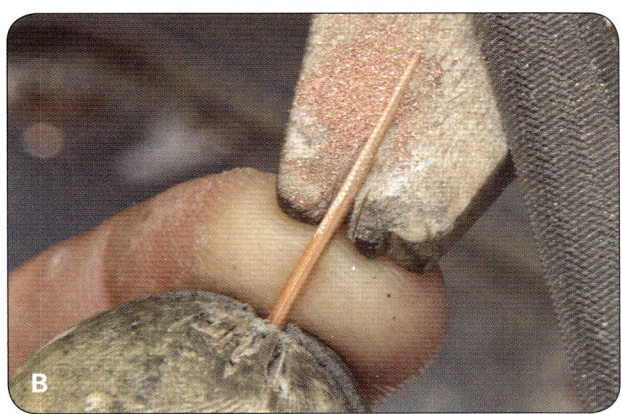

Solder the jump rings

Solder eight 4mm and two 5mm 18-gauge sterling silver jump rings with silver hard wire solder (see *Soldering Techniques*, p. 26) **[A]**: Place the jump rings about 1" (25.5mm) apart on the solder board. The cold board will slow down the heat and prevent accidental melting, which is helpful if you're new to soldering. It is also cleaner than using charcoal or firebrick, which can add dust to the flux, obscuring the join. After you have more experience and control, you can use charcoal or honeycomb boards to speed up soldering. Solder, quench, and then pickle all of the jump rings for 10–15 minutes.

Taper and form wires

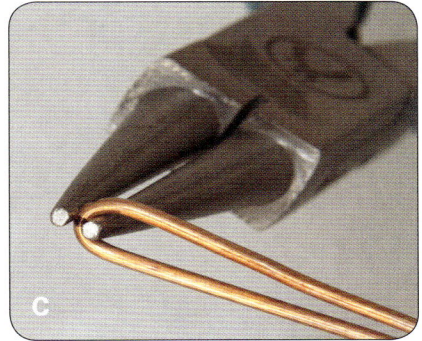

Cut two 3" (76mm) pieces of 16-gauge round copper wire. Each end should be flush. Hold one of the wires with a ring clamp with about 1" (25.5mm) exposed. File a shallow notch into the edge of one leg of the bench pin. Rest the part you're filing in the notch on the pin. Support the rest of the wire with a finger to keep it from bending.

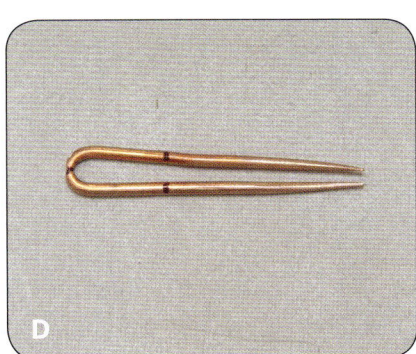

File a ½" (13mm) taper with the flat side of a cut 0 half-round file **[B]**: Start by filing four flat sides, making the last half inch square. This will make it easier to track your taper as you work. Then file the tip to a point, alternating sides by turning it as you file. As the taper forms, file further back on each side, making a smooth transition from the point to the original gauge. Finally, file off the corners of the square, and blend the shape back to a tapered round wire. Use a cut 4 file to remove file marks, and sand it smooth with sanding sticks, first 320- and then 600-grit. Removing the scratches will make it easier to polish later (see *Sanding*, p. 103). Repeat for both ends of the two wires.

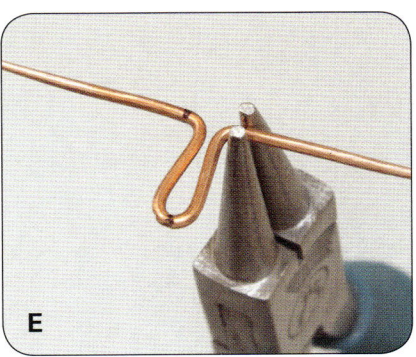

To form them, measure and mark the center of each wire with a permanent marker. Grasp on that mark with the narrow tip of roundnose pliers, and bend the ends towards each other to form a long loop **[C]**. Mark a second measurement 12mm (½") down from the loop on each leg **[D]**. Grasp with the tips of the roundnose pliers again, and bend each leg 90 degrees **[E]**. Repeat to shape the second wire. Flatten the shapes with a rawhide mallet on a steel block.

Solder the jump rings to the copper wires

To make it easier to find them later, mark each join with a permanent marker. Use a charcoal block or honeycomb to solder the jump rings to the copper wires. The surface should be flat to make aligning the pieces easier.

Use the sticky nature of the flux to help hold the parts together for soldering: First, paint a thin coat of flux on the surface of the charcoal where the parts will be arranged. This will create a base to bond the parts to the charcoal when the flux cools and hardens. Flux all the jump rings and wires as usual.

Arrange the jump rings on the flux spot on the charcoal, with the 5mm ring in the center of the bottom row. Arrange all the joins toward the copper wire, to hide the joins under the silver balls that will be added later. Cut and flux eight 1mm (or smaller) chips of medium silver solder (using small chips of solder will limit the overflow of extra solder onto the copper).

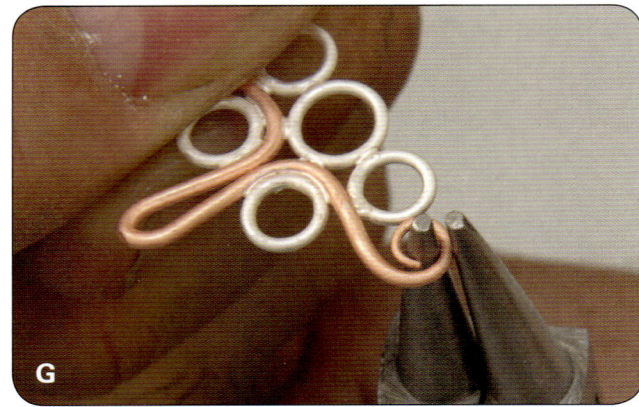

Heat the rings and wire until all of the flux, including what's on the charcoal, turns clear. Move the flame to the solder, and scoop up the first ball with your pick. With a light flame, place the ball on the join between the copper and a ring **[F]**. Continue heating that section until the solder flows as the metal turns a medium red. As the solder flows, shift the flame to help it flow into the join and away from the copper. Repeat for each join where the rings touch the wire, and solder each lower 4mm ring to the 5mm ring. Repeat for the second earring. Quench and pickle for 10 minutes.

tip

Flux can be annoying or helpful, depending on how you use it. If flux is too cold, solder will fall off and nothing will move. When the flux is liquid and warm, it holds the solder and parts can be shifted. Heating the flux first and letting it cool into a hard, glassy glaze while you scoop up the solder helps to keep everything from drifting apart. Once you start, don't turn the torch off! You have limited time to scoop the

solder and place it before the flux cools so much that the solder will fall off. Even so, it will take a light amount of heat to bring the flux up to its sticky liquid state so that it will grab the solder. Too much heat, and the flux will loosen and the parts can move. If they do, just simmer the flux and use the pick to put them back. During this whole time, the metal should look clean and matte, with no sign of even the faintest red glow.

After pickling, rinse and dry the earrings. Use the roundnose pliers to roll the ends of the copper wires into spirals. Make each spiral touch the jump ring next to it **[G]**. Flux all sides again and solder the spirals to rings with 1mm chips of medium silver solder. Pickle again.

Melt balls and solder
The earrings are decorated with small sterling balls made by melting pieces of wire. Cut four 6mm (¼") and two 12mm (½") pieces of 18-gauge sterling round wire. Spread the pieces on the charcoal and melt each piece into a ball **[H]**: Keep the tip of the cone inside the flame near the silver until it turns glossy

and pulls up into a ball. Stand by with a pick in case one of them tries to roll off. Allow them to cool to a dull red and then quench. If you don't like a shape, melt it again. The bottom of each ball should be silver, because contact with the charcoal keeps it clean. It's OK if the top remains black with scale, because the flat bottom will be soldered to the earrings. If there is charcoal on the bottom, remove it after it's cool or pickle it.

tip

Why is the bottom of the ball flat? When the silver balled up, the bottom conformed to the flat surface of the charcoal. The larger the ball, the wider the flat bottom will be. If you want the bottom to be round, make a shallow depression in the charcoal with a round dapping punch that is a little larger than the ball. The molten ball will try to match the shape.

Collect all of the balls on the solder board. Flux each one on all sides. Cut and flux six 1mm chips of easy silver solder. The balls will be sweat-soldered onto the earrings: Solder will first be melted onto each ball by scooping up a small chip, placing it on the flat bottom of the ball, and heating until it melts into a flat puddle **[I]**. Then the balls will be placed solder-side down on the fluxed earrings, with one larger ball in the middle, near the loop. When the earrings are heated on the charcoal to solder temp, the solder will flow between them **[J]**. It can be hard to see the solder flow under the balls. Look for signs including the flux turning super clear, the metal glowing a light red, and the balls sinking slightly. Also, the medium solder between the rings may ripple a little, like little rivers of mercury. Pickle for 10–15 minutes. If they didn't join, they will come off in the pickle, so use a plastic sieve to hold the earrings during cleaning (see *Six simple steps to soldering: Pickle*, p. 20).If they come off, just reflux all the parts, set the ball in place solder-side down, and try again. If you can see solder on the bottom of the ball, it's not necessary to add more.

Sweat-soldering like this is a great way to control where the solder flows. If chips of solder were placed next to the silver balls on the earring, they could just as easily flow away from the join and all over the copper.

Polish and add patina and beads

Polish using the recommendations for abrasive attachments below **[K]**. To finish, add patina with liver of sulfur (see *Adding Patina*, p. 107), and bring back the highlights with a polishing pad. Add a beaded-headpin dangle to each of the bottom jump rings using wrapped loops.

POLISHING ABRASIVE	WHERE TO POLISH
Mini half-round file, #2 cut, or gray polishing pin	Clean up excess lumps of solder on the jump rings and wire.
Polishing pins: gray, brown, green	Inside the rings and loops.
Radial bristle disks: white, red, blue, peach, green	Remove scale and scratches and polish through all grits to a mirror finish. Alternatively, stop at blue (400-grit) and tumble in stainless steel shot for 2 hours (see *Burnishing with a Tumbler*, p. 106).

DOG TAG CHAIN

The fluid motion of chain links is so lovely. Making your own chain of mixed metals with soldered links is a great way to add personal style to jewelry and excellent practice for closing links without accidentally soldering the rest into a jumble. Repetition with simple exercises, like soldering multiple links, builds long-term skills.

LEVEL
Beginner

TECHNIQUES
Soldering chain
Filing and using shears
Using a third hand
Making headpins

MATERIALS
- **10** 8mm ID 18-gauge sterling jump rings
- **17** 4mm ID 18-gauge sterling jump rings
- 2mm ID 20-gauge sterling jump ring
- 1" 16-gauge sterling silver round wire, dead soft
- ¼x3" (6x75mm) 24-gauge copper or brass sheet
- Easy and hard silver solder

TORCH
Micro butane torch, maximum flame, or small torch with #5 tip, medium flame

FLUX
Paste flux

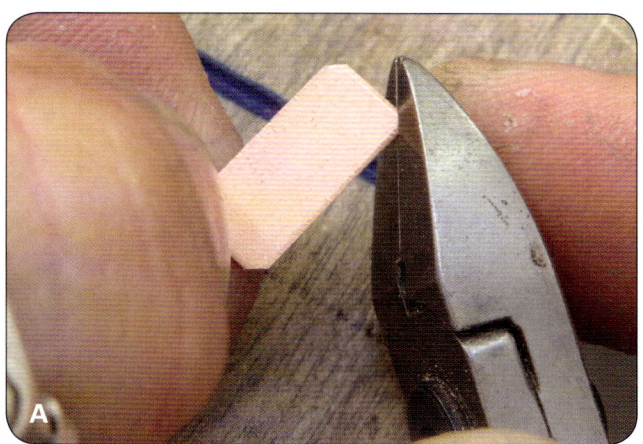

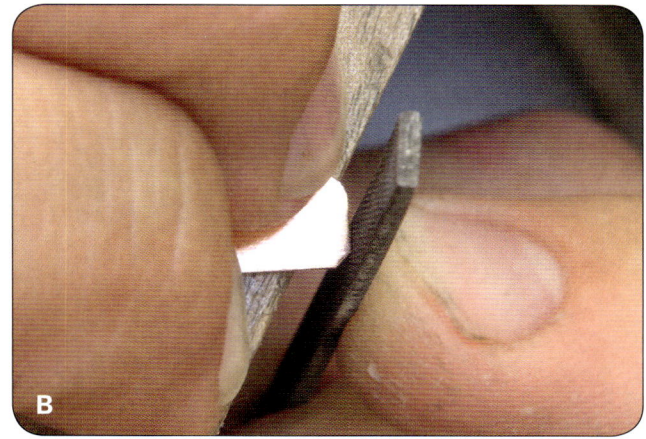

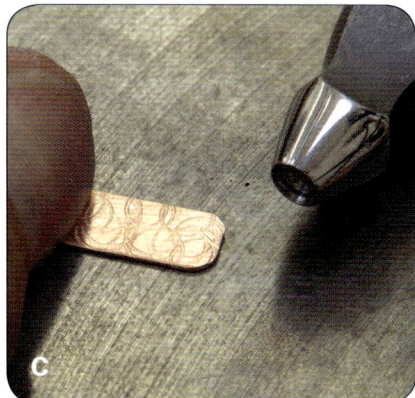

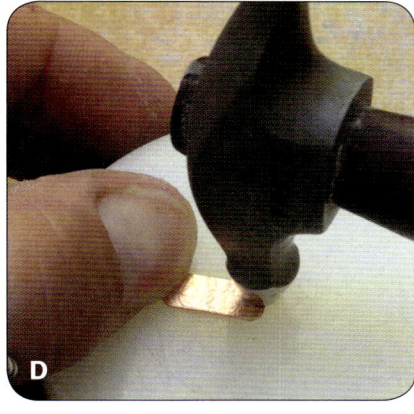

Cut, texture, and punch copper links

Measure and cut the copper sheet with shears into five 6x15mm (¼ x ½") pieces. The sides should be straight and at right angles. To round the corners, first trim them equally at 45 degrees with shears or flush cutters **[A]**. File off the sharp corners to round them **[B]**. Texture each link with texture hammers or stamps on a steel block **[C]**. For two-sided texture, switch to a hard nylon block **[D]**. This will allow you to texture without smashing the pattern on the first side, but the stamps or hammer marks can't be too subtle or they won't make an impression with the plastic block. Make a hole on both ends of each link with a screw-down metal punch **[E]**. Remove any burrs with a sharp ⅛" drill bit, turning it in the hole to scrape the edge lightly **[F]**. Using the drill bit prevents any file marks or scratches.

Solder jump rings

Normally, when you solder a custom chain, solder as many jump rings closed on the board as possible before soldering them together as links. For example, for a simple chain, solder half of the jump rings closed on the board, and then join each closed pair of links with an open jump ring that will be soldered later.

For this project, solder one 6mm and all of the 8mm 18-gauge sterling jump rings closed with 1mm chips of hard silver solder (see *Soldering Techniques*, p. 26) **[G]**. I used a honeycomb board to heat them fast. The clean ceramic surface leaves no extra dust to clean up. For even less cleanup later, split the solder balls in half for just the right amount to fill the join (see *Splitting solder*, p. 29). Quench the rings, but don't pickle them. Check the joins before assembling them into the chain.

Make a toggle

The toggle is made by balling up wire ends and flattening them with a hammer. Cut 25mm (1") of 16-gauge round sterling wire. Flux the entire wire before heating. Hold one end with tweezers. Keep the wire vertical and heat the lower end until it balls up **[H]**. With a butane torch, adjust the flame to an oxidizing flame (see *Adjusting the Flame*, p. 24). Place the sharp tip of the cone near the end of the wire, parallel to the solder board, to ball it up. If you're using a small torch, also use an oxidizing flame, but don't get as close. Adjust the angle of the wire while the ball is molten and remove the heat when it's centered. Repeat if necessary to reposition or adjust the roundness of the ball. Repeat on the other end to make a ball of equal size. Quench and pickle for 2 minutes to get the flux off. Hammer both ends flat with a polished planishing hammer on a steel block.

Close a 2mm 20-gauge sterling jump ring with a 1mm chip of hard silver solder. Pickle for 2 minutes. File a small flat spot at the join for a better connection to the toggle bar **[I]**. Use the flat side of a cut 2 needle file. Hold the jump ring with flatnose pliers and brace the pliers against the bench pin for better control.

Flux the toggle and small jump ring, placing them together on the honeycomb board. Solder with a 1mm chip of sterling hard solder. Pickle for 10 minutes.

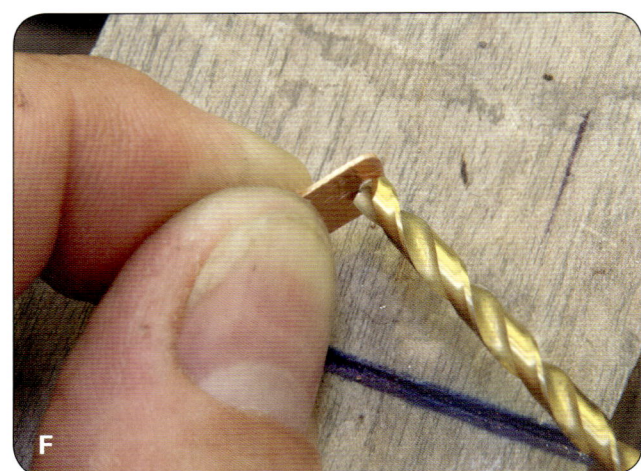

Assemble the chain

Collect the toggle, the copper dog tags, the closed 8mm rings, and the rest of the 4mm jump rings, and assemble the chain, connecting everything with 4mm sterling 18-gauge jump rings. Connect the toggle with a 4mm ring to a closed 4mm ring, and use another ring to connect it to a dog tag link. When you're satisfied with the pattern of the chain, use bent and chainnose pliers to close all the open rings for soldering **[J]**.

Solder the links

If you lay a link on the solder board, it's not only hard to keep it in place while you solder, but the linked rings are easy to overheat and accidentally solder together into a frozen mess. Instead, hold the other links back with a third hand, resting the join to solder on the edge of a honeycomb block. The link to solder should be loose and not held with the cold steel tweezers, which would make it take longer to solder. Resting it like that makes it more stable for placing the solder ball, and helps it heat

faster. Flux the link to solder and the links inside the jaws of the third hand. Scoop up a 1mm ball of hard solder, and as you warm the flux to a clear glaze, place it on the join **[K]**. The flame should point away from the other links at all times, so angling it over the third hand works best. Don't be shy with the hottest part of the flame, whether it's butane or oxy/propane: Move

in close to the ring and solder it quickly. This will give the heat less time to damage the other links. Quench the link and continue to solder the others until the chain is complete. Pickle for 10 minutes and check your joins.

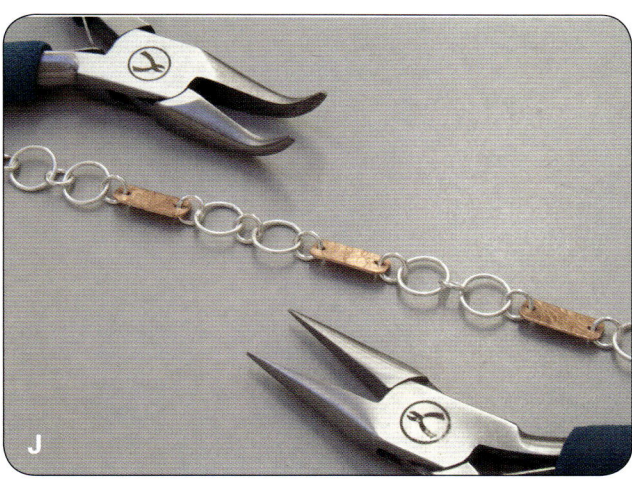

Polish the chain

Polish the chain links using the abrasive attachments recommended in the polishing chart below. Use caution; polishing loose chain can be dangerous. If the chain gets caught in the flex shaft or any spinning parts, it can catch or break your fingers. Expose only one link at a time to polish, keeping the rest tucked away safely inside your hand. As your soldering skills improve, the flux will protect the metal and you'll have very little scale to clean up.

POLISHING ABRASIVE	WHERE TO POLISH
Silicone polishing wheel, black medium	Remove lumps of solder from the links.
Radial bristle disks: red, blue, peach, light green	Remove scale and scratches and polish from red (220-grit) through light green to a mirror finish. Alternatively, stop at blue (400-grit) and tumble-polish in stainless steel shot for 2 hours.

There is no vanity in jewelry making. At some point, everyone needs magnification to see more clearly. Using low-power (1–2x) magnifiers to help you see more clearly at a safe distance of at least 10" (25.4cm) makes it easier to find and solder those small joins. Headband magnifiers or clip-ons that can sit on your safety glasses come in a wide range of magnifications to suit your eyesight and the job required.

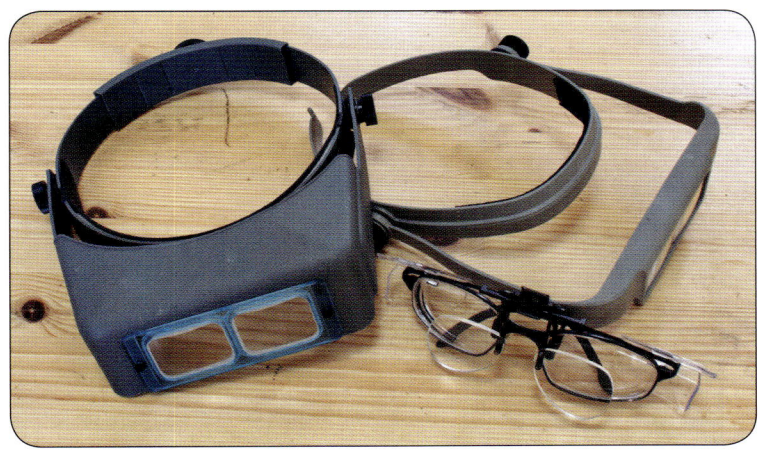

BEADS AND RINGS EARRINGS

What could be more amazing than soldering jump rings with beads attached? Most beads will shatter, burn, or break under the intense heat of soldering. With the techniques shown in this project, you can safely solder near most beads without any damage. It's not a bad idea to practice this project with copper jump rings and less expensive beads before you try sterling and these beautiful glass donuts.

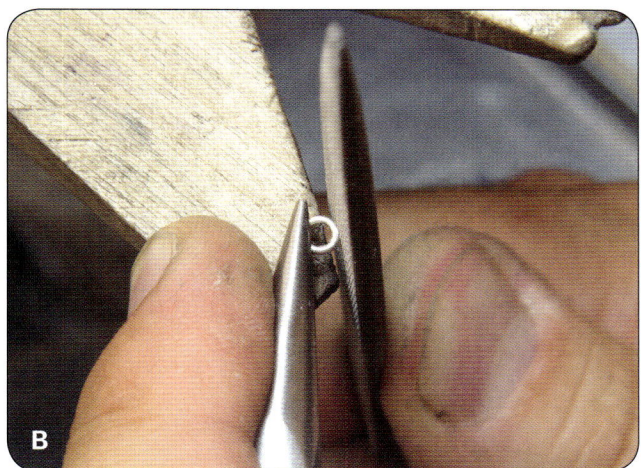

The beads for this project should be durable enough to withstand a little heat (see *Soldering Near Stones*, p. 31). They should fit loosely on the jump rings and be small or wide enough to leave space between the bead and the join. Some materials are too delicate to endure the heat of soldering. Don't leave delicate materials in the pickle for long. Even glass beads can be etched by the acidic pickle.

Assemble earrings

Assemble the earrings, connecting the large glass donuts with 11mm sterling jump rings. Connect the smaller glass bead to the large donut with the 8mm ring. Close all the joins for soldering (see *Closing jump rings for soldering*, p. 26).

Solder rings

Using the third hand as shown in *Dog Tag Chain*, p. 42, hold the beads and other links away from the ring to be soldered, and rest the join on the charcoal block **[A]**. Since these are large rings, it's OK to use just the tips of the tweezers to hold the jump ring on the opposite side from the join. Keep the edges of the beads away from the charcoal, since it can reflect a lot of heat. Flux the entire link. Angle a sharp oxidizing flame away from the beads and toward the join. When the flux clears, place a 1mm or smaller ball of easy silver solder on the join. Solder as quickly as possible to avoid overheating the glass. Use your tweezers to move to the next ring, and don't quench! The glass could shatter if cooled too quickly. Repeat for all rings on both earrings. Allow the earrings to air cool, and don't pickle.

Solder small rings

Solder a 2mm 20-gauge sterling jump ring closed with a 1mm or smaller chip of easy silver solder. Pickle for 2 minutes. Hold the ring with chainnose pliers, and file a small flat spot at the join using the flat side of a file **[B]**. Flux and set up the 11mm jump ring at the top of the earring soldered earlier, with the join resting on a flat spot on the charcoal block. Flux the small ring and put the flat spot against the large one. Solder with a 1mm ball of easy silver solder

[C]. Let it air cool. Repeat for the other earring. Pickle both pieces. Since you used only a small amount of flux, they should pickle in less than 2 minutes. Remove and rinse.

Polish

Polish using the abrasive attachments recommended below. Because you used a lower-temp solder and carefully protected the silver with flux, any firescale should be minimal and easily removed.

POLISHING ABRASIVE	WHERE TO POLISH
Silicone polishing wheel, black medium	Remove lumps of solder from the jump rings.
Radial bristle disks: red, blue, peach, light green	Remove scale and scratches and polish from red through light green to a mirror finish. Don't tumble-polish; beads could be damaged.

Keep loose links and beads tucked inside your hand and away from the power tool, exposing just the link you're polishing to the bit. With the large rings, avoid reaching too far inside the ring with the bit or it can catch and pull it onto the flex shaft. If anything like that happens, let go of the chain and turn off the tool immediately.

FREEFORM PRONGS

Create settings for metal stampings, beads, unusual stones, and river rocks with a simple framework of wire prongs. The basic concept is to create a simple frame or base that fits within the outline of your piece to set. Solder on wire prongs to later bend over the edges, capturing it against the base.

LEVEL
Beginner

TECHNIQUES
Soldering wire

Designing a custom prong setting

Setting prongs

MATERIALS
- 4mm ID 18-gauge sterling silver, copper, or brass jump ring
- 12" (30.5cm) 18- or 16-gauge sterling silver, copper, or brass round wire, dead soft
- Easy, medium, or hard silver solder wire
- Something to set

TORCH
Micro butane torch, maximum flame, or small torch with #5 tip, medium flame

FLUX
Paste flux

A

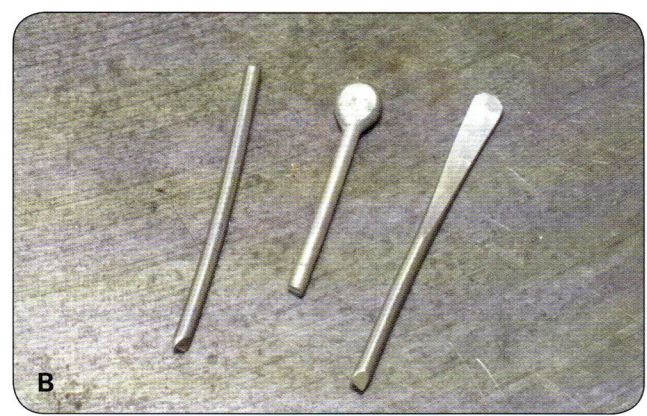

B

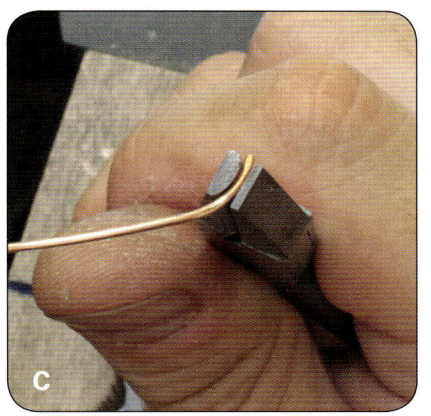

C

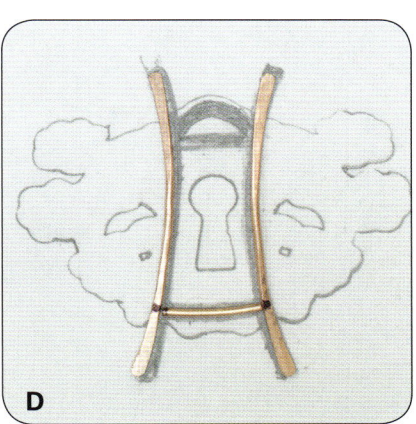

D

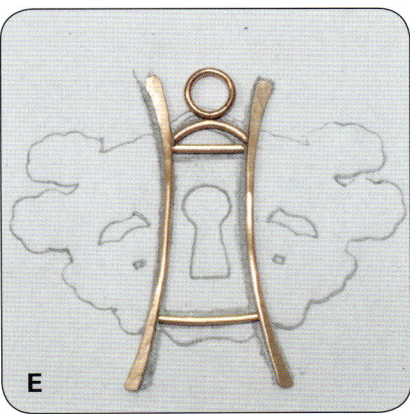

E

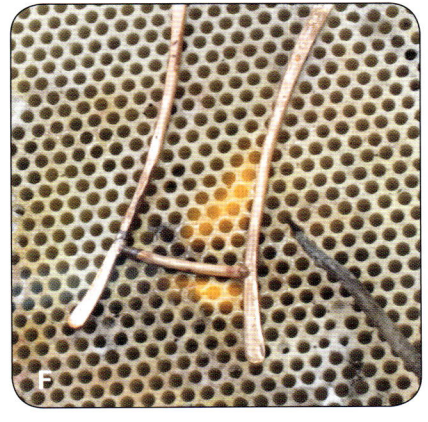

F

Sketch the setting

Draw an outline around your piece to set (in my case, it was a metal stamping). Be sure to outline any openings too. Sketch ideas for the frame and prongs. In this example, the frame extends to create four prongs **[A]**. Look for good spots to place the prongs, like in the notches on the edge of this stamping. To match my stamping, I made the frame and prongs out of copper (the patina I added later hid the prongs well). Using a contrasting metal, like sterling silver, is another option that would have accented my setting. The ends of the prongs can be varied: flattened headpins, forged wedges, or straight wire **[B]**.

How long should the prongs be?

The prongs need to be able to bend over the edges of your piece, with enough wire left over to hold your piece securely against the base. Too long, and they can obscure the beauty of what you're setting. To measure for the prongs, use something flexible and cheap, like leftover twist-ties from bread packaging or strips of paper. Bend the scrap around the edge of the piece like a prong, mark where it would meet the frame

underneath, and then measure that against the ruler. In this example, the prongs extend 5mm past the point where they stretch away from the outline of the frame. Except for flattened headpin-style prongs, prongs can be trimmed to size during setting, so make them a little longer than needed.

Make the frame and prongs

I cut the long wires from 18-gauge copper wire, 2mm shorter than the drawing. Use your fingers or flat/half-round forming pliers to form the wires to match your sketch **[C]**. I forged the ends into wedges with the polished flat face of a goldsmith's hammer. As they were forged, I compared them with the drawing until they matched **[D]**. Don't thin the ends too much; they need to be thick enough to make strong prongs. I flush-cut the rest of the parts, including the arch and the straight pieces, from the same copper wire and filed the ends with a flat needle file to create tight-fitting joins and positioned the jump ring at the top **[E]**.

Solder the parts together

It's easiest to solder one piece at a time. Mark with permanent marker where you want to attach the first piece to the long prongs. Follow the six simple steps to soldering: Flux the pieces completely and arrange them on a honeycomb board for a flat, clean surface. Cut and flux several 1mm chips of silver solder (easy, medium, or hard). Make sure the join is flush. Heat the entire piece until the flux turns clear. Simmer the heat as you place a ball of solder on the join. Continue heating to solder it together. Repeat for the other end **[F]**.

Allow the piece to air cool to avoid cracking off any flux in the water, then mark it again for the next part. Add a light coat of flux on the pieces soldered so far, and flux the new piece. Solder the short bar across the top with the same grade of solder. Place and solder the arched piece. Solder the 4mm 18-gauge jump ring closed separately, file a small flat spot at the join for a better fit to the frame, and solder the ring to the setting **[G]**. Pickle for 15 minutes.

Set the prongs

Polish and add patina to the setting before positioning the piece: Dip it in liver of sulfur for 2 minutes (see *Adding Patina*, p. 107). Lighten the patina with a polishing pad to show a little bit of the copper. Buff to a bright finish using the abrasive attachments recommended below.

Center the setting under the stamping. With your finger or roundnose pliers, bend the prongs 90 degrees so they will cradle the stamping **[H, I]**. Don't bend the prongs at the solder joins—they will break.

Rest the setting and stamping on soft leather to prevent scratches. To protect the finish on the prongs and the stamping, use soft materials for prong pushers, like wooden chopsticks or the eraser end of a pencil. Press prongs down in pairs of opposites. Place the end of the pusher

on the prong and press it over until it almost touches the stamping. Stopping short will give you enough room to adjust the fit and to keep it centered. Go to the opposite prong across from it, and press it over the same way. Repeat for the last pair.

If the prongs look long, trim them with flush cutters and use a mini file or a blue silicone polishing wheel to round off the corners. Brush away any dust and touch up any missing patina by applying solution directly with a natural bristle brush.

 tip

Applying hot liver of sulfur directly to a piece with a brush takes a little longer, but it's more precise.

Neutralize the patina with cold water. Press down each pair of opposite prongs until they're flush to the metal of the stamping **[J]**. The piece should be secure in the setting. The back of the setting should look nice too, with the frame fitting flat and tight.

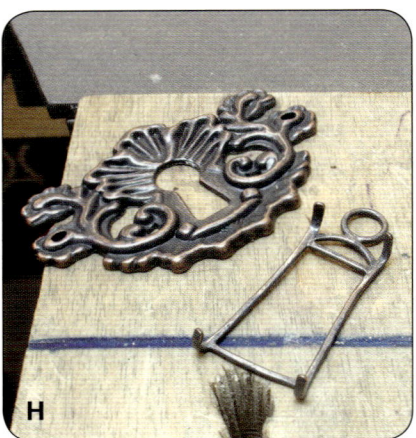

POLISHING ABRASIVE	WHERE TO POLISH
Silicone polishing wheel, black medium	Remove lumps of solder from the wire.
Radial bristle disks: red, blue, peach, light green	Remove scale and scratches and polish from red (220-grit) through light green for a mirror finish. Alternatively, stop at blue (400-grit) and tumble in stainless steel shot for 2 hours.

LOOKING SHARP BEZEL

Most jewelry students want to get to setting stones as quickly as possible, because they are so beautiful and the process is so satisfying. A sharp-cornered stone shouldn't be your very first bezel-setting project; it's better to get some practice with a round, oval, or irregular cabochon first. Then, tackling the fine adjustments required for a cabochon with sharp corners like this one will be far less challenging.

LEVEL
Beginner

TECHNIQUES
Making a bezel setting

Adjusting a setting for sharp corners

Making a bail

MATERIALS
- 1½x1½" (40x40mm) 22-gauge sterling silver sheet (varies with cabochon size)
- 12" (30.5cm) 28-gauge fine silver plain bezel strip, ⅛–¼" (3–6mm) (varies with cabochon size)
- 4mm ID 18-gauge sterling silver jump ring
- Easy, medium, and hard silver solder
- Cabochon

TORCH
Large flame-butane torch, medium to maximum flame, or small torch with melting tip, medium flame

FLUX
Cupronil and paste flux

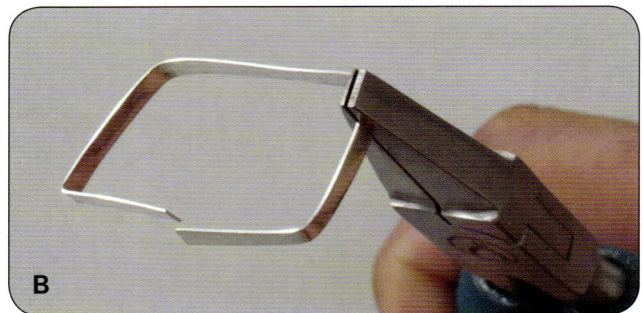

Form and solder the bezel

Trim one end of the bezel strip at a 90-degree angle with flush cutters or shears. Take your time to line up the cutters before trimming. A straight cut will require less filing. File off any bevel. Using razor flush cutters saves time by not making beveled edges. With the flat back of the stone facing you, start in the center of the longest side and wrap the bezel around until it overlaps. Pull tight on the bezel for a snug fit, but don't crush the soft fine silver. The bezel walls must stay straight so the stone can pass through it later. Mark the overlap with a scribe **[A]**. Make a second 90-degree flush-cut. For any sharp corners, make a crisp bend with flatnose pliers **[B]**.

Press the bezel around the girdle of the stone again and check the fit at the join. There should be no gaps between the girdle and the bezel. The join can have up to a 1mm gap for stones without corners and still be stretched to fit after soldering. But for a sharp-corner bezel, the fit should be as exact as possible. Close the join by overlapping the ends repeatedly, creating tension until they stay together. Line up the corners. Use flatnose or flat/half-round pliers to align the join so that the ends are parallel. Don't twist the bezel to make it fit; it has to be perfectly flat for soldering to a base later.

Hold the join in front of a fluorescent light and look for any gaps. If the angles are off, use a flat needle file to correct it **[C]**. Open it just wide enough for the file to fit. One side of the join will guide your filing. Work on a bench pin to steady yourself, but hold the end you're filing firmly. Use only the last inch of the file filing in short, controlled push strokes. Be careful not to round the corners! Using a long file stroke, letting the metal wiggle, or filing back and forth can make a poor join. Check the fit around the stone again, in case it's too short after filing.

The bezel is fine silver and doesn't contain any copper to make firescale. Flux just the join with paste flux. Place it on a flat surface, like a honeycomb board. For an easy join, forge flat some hard silver solder wire or use sheet solder, and cut a 1mm chip. Flux it lightly and place it under the join of the bezel. Watch the join; it can pop open during setup and soldering. If it opens or overlaps, stop and fix it. With a medium flame, preheat the bezel until the flux turns clear and then focus the heat down the center of the join until the hard solder flows to the top **[D]**. If the bezel starts to glow brighter than red or turn glossy, pull the heat away immediately or it will melt. Pickle for 5 minutes. Inspect the join with a loupe and make sure it's completely soldered.

Solder the bezel to the base

Before soldering the bezel to a base, recheck the fit with the stone. Rest the bezel on a flat surface. The stone should pass easily through it from the top. On the backside, the bezel should fit against the stone with no gaps. If it's too tight, it can be stretched. If there is a lump of solder on the inside, remove it with a white silicone polishing wheel before checking the fit.

For bezels without corners, use a round mandrel, like a ring mandrel to stretch it. Bezels with corners may require a square mandrel to work on the straight sides. Start first with a rawhide mallet to tap out any wrinkles. Check the fit again. If it's still too tight, use the polished flat face of a goldsmith's hammer to planish it. Start at the join. Hammer a couple of rows up and down the height of the bezel. Check the fit on the stone often, because fine silver stretches quickly. If the bezel is too tight and needs to flex to fit the stone, it won't fit after being soldered to a base.

The bezel needs to lay flat against the sheet metal for soldering. You can correct any slight twisting with a nylon block. Place the bezel

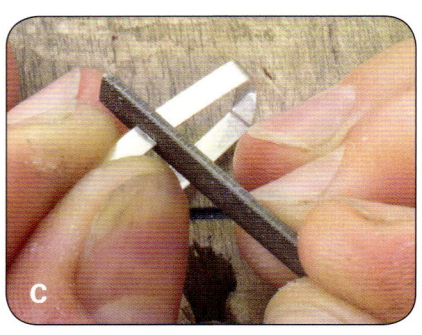

on a steel block with nylon block on top. Tap the nylon block with a rawhide mallet **[E]**. This should help it lay flat for sanding. Hold a piece of 320-grit sandpaper on a flat, clean surface and sand the top and bottom of the bezel. Check the edge by resting it on the steel block. The bezel should lay flat against the steel with no gaps. Keep track of which side is the top; most stones only fit in one direction. After sanding, you can stain the top edge with a permanent marker so that you don't solder it upside-down.

Measure a piece of 22-gauge sheet metal that is 3mm larger than the bezel on all sides. Open a pair of dividers to this measurement, and drag one leg against the straight edge, while pressing down on the other leg, scribing a parallel line **[F]**. Use these lines for guides to shear or saw it to size. Using less sheet metal will make it faster to heat for soldering. Soldering the bezel to a slightly larger base is easier than trying to fit it to an exact shape. Afterward, it can be trimmed to make a border.

If you want to stamp the base with your logo or a quality stamp, like .925, do it now before soldering **[G]**. Flatten the sheet with a rawhide mallet. If the bezel doesn't lie flat against it, anneal it and/or sand the sheet on 320-grit paper until it fits. Clean the sheet metal before soldering to remove any oils or dirt, which could repel the flux (see *Six simple steps to soldering: Clean*, p. 19)

When all the solder is in place, turn up the flame to max on the butane torch or to medium with a melting tip, and keep heating from underneath. When the flux takes on a slight green cast, the medium solder should flow. If it's reluctant, bring the heat to the top and aim it at 45 degrees along the join line, turning the mesh to follow the solder as it flows around. Be careful not to overheat the bezel! Pickle for 15 minutes.

After soldering, check the joins and the fit of the stone, placing some dental floss underneath the stone first **[I]** The floss allows you to pop the stone out again. If it's too tight to fit, then the bezel will have to be unsoldered and stretched (see *Unsoldering*, p. 30).

Trim the base

Draw the outline of your border around the stone. For a parallel border, set the dividers to the width you want and trace the base of the bezel, pressing the other leg down to scribe a line **[J]**. I recommend sawing the base with a #2 blade. If you use shears, make short, straight cuts. Shearing can warp the metal. Rest the base on a bench pin with the edge overhanging the side, and file to the line with a cut 2 hand file. Then file along the edge to refine the shape **[K]**. Remove any burrs by running a scraper along the edge.

Cut enough 2mm pieces of medium silver solder to be placed every ½" (13mm) around the bezel. Dab each piece lightly with paste flux to make it easier to pick up. Check the fit of the stone inside the bezel one last time before soldering. If it's even slightly off, it won't fit!

Flux the sheet metal and bezel with Cupronil to stop firescale (see *Using spray flux*, p. 22). Apply Cupronil to both sides of the sheet first, placing it face up on a soldering tripod. Flux the join on the bezel and put it on the sheet. The base is thick compared to the thin bezel, so heating the base first helps to draw the solder into the join. Heating from above can spread the solder all

over the sides of the bezel or melt it. Use a medium flame to warm the flux from underneath the screen **[H]**. A larger flame is required to heat the screen and metal to soldering temperature. A small micro butane torch won't work, except on small settings. When the flux liquifies and the bezel settles against the sheet, simmer the heat as you place the solder around the outside of the seam. Start with one piece at the join in the bezel and place a chip every ½" (13mm). It's easier to clean up any extra solder around the outside. Any lumps on the inside are harder to remove and could affect the fit of the stone.

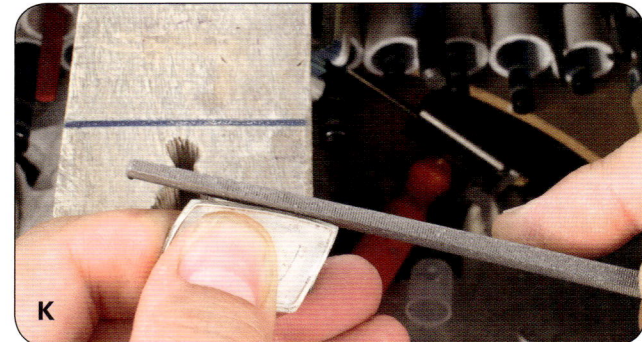

Solder on a jump ring

Use paste flux or Cupronil to flux a 4mm jump ring. Close it with hard silver solder. If it's still free of scale, don't pickle it.

Draw a line where you want to solder the ring on the back of the setting with a permanent marker. Apply Cupronil to all sides of the setting. Place it bezel-side down on the honeycomb board. Hold the ring in place with a third hand, with the join over the base. Leave at least half of the ring past the edge, so that the bail will fit inside. Solder with a 1mm ball of easy solder. Use a maximum butane flame or a medium flame with the melting tip [L]. Heat mostly the setting, away from the ring, to balance the heat. Quench. Pickle for 10 minutes.

Make and solder on a bail

A simple bail is a V-shaped loop with a narrow point. It's an easy finding to make. Start with a 6x25mm (¼x1") strip of 22-gauge sterling sheet. Divide the strip in half with scribed lines along the length and width. The blank for the bail looks like a long diamond shape. Use a permanent marker to connect the ends of the centerlines to make the diamond shape [M]. Cut with shears. File the sides to refine the shape.

To form the shape, hold the center of the bail with roundnose pliers. Bend the ends halfway around the pliers. Flip the bail over and finish bending. This will keep it from tilting, since the jaws of the pliers are tapered. Insert the bail into the jump ring on the setting to check the fit. The join should be far enough away from it to solder easily. To bring the ends together for soldering, use chainnose pliers to gently curve them toward each other [N].

Apply Cupronil to the setting and bail for soldering. Point the join away from the setting. Angle a medium flame inside the bail and away from the setting, evenly heating both sides [O]. Place a 1mm ball of easy solder on the clear flux, and continue heating to draw it into the join. If you do this correctly, the bail should solder before the flux on the setting has a chance to clear. Quench. Pickle for 10 minutes.

Polish and add patina

Polish and add patina to the bezel before setting, because stones can be scratched by polishing abrasives. Use the abrasive attachments recommended on the next page. Polish the inside of the bezel only if the stone isn't opaque. Antique just the areas you want to darken and bring back the highlights with a polishing pad.

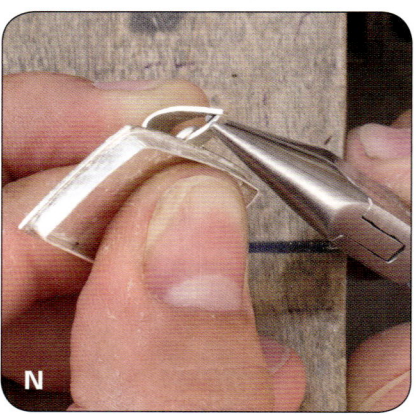

POLISHING ABRASIVE	WHERE TO POLISH
Silicone polishing wheel, black medium	Remove join lines and excess solder on the bezel and bail. Remove file marks on the border.
Split mandrels: 320-, 600-, 1200-grit	The flat sheet-metal back of the setting.
Polishing pins: gray, brown, green	Inside the bail and jump ring.
Radial bristle disks: blue, peach, light green	The entire setting and bail.

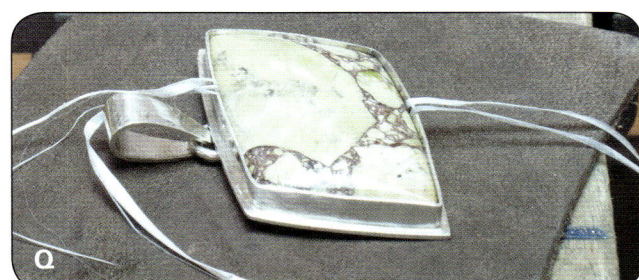

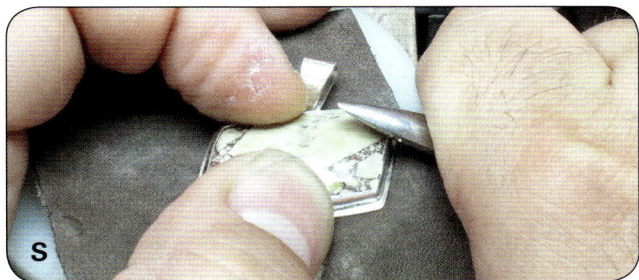

Set the bezel

Clean the stone and setting. Rest the piece on leather scrap to protect the finish. Insert the stone over dental floss and check the height of the bezel [P]. This stone is too low, with too much bezel height, especially at the corners. You can file or sand the top edge to the correct height or raise the cabochon with some pieces of plastic yogurt lids cut to fit inside the bezel. Two pieces raised my stone to the right height. Use a flat #2 needle file or a blue silicone wheel to lower the corners by a few millimeters (no lower than half the bezel height) [Q]. If the corners are too high, they're hard to push all the way down.

Use a prong pusher to crimp the raised edge of the bezel against the stone. Start with the corners, and press them down in pairs of opposites [R]. Working around the stone in a circle will leave hard puckers that are difficult to crimp. Continue crimping in pairs of opposites until the bezel is flat against the stone. For a smooth, polished edge, use a burnisher [S]. Hold it like a vegetable peeler and rub back and forth repeatedly with pressure to bevel the bezel to a fine edge. Burnish away from corners to avoid stretching them up from the stone. You may see a few lumps or puckers left at the corners [T]. The lumps can be thinned using a blue silicone wheel [U]. Although most stones are not scratched by the fine blue wheel, turn it so it doesn't touch the stone. Be careful not to break through the thin bezel. Repeat the burnishing to finish smoothing the bezel after polishing [V].

DOUBLE-SIDED BEZEL

Not every stone has a flat back like a cabochon.
I bet you have many interesting objects you want to
bezel-set—things like river rocks, beach glass, and tumbled
pottery. This variation on a traditional setting has a bezel that
wraps just around the edge of the stone, showing off both sides
of the piece. The setting has to fit snugly around the stone, and you
must have enough bezel to push down around both sides for a secure fit.

LEVEL
Beginner

TECHNIQUES
Making a backless bezel
setting around just the
edge of the stone

Soldering on a bail after
setting the stone

MATERIALS
- 12" (30.5cm) 28-gauge fine-silver plain
  bezel strip, ⅛–¼" (3–6mm) (varies with
  size of item to set)
- 4mm ID 18-gauge sterling jump ring
- 6x25mm (¼ x1") 22-gauge sterling
  silver sheet
- Easy and hard silver solder
- Two-part clear epoxy
- Beach glass, river rock, or similar item
  to set

TORCH
Micro-butane torch, maximum flame, or
small torch with #5 tip, medium flame

FLUX
Paste flux

What can you set in a double-sided bezel?

This technique is made for setting pieces that are rounded on both sides, which will give you that all-important taper to set the bezel against. I don't recommend using this method to set normal cabochons, because the flat side requires pushing the bezel over 90 degrees. Sharp angles can leave hard wrinkles that can't be set. Instead, set them with a traditional bezel, but make an open back by piercing out the inside or adding a soldered ledge.

The ideal first piece to set for this project has a straight girdle. In other words, it isn't twisted or warped like a potato chip, which is harder to wrap the bezel around. Also, the more evenly rounded the edge of the girdle, the easier the setting. Pieces with randomly angled edges and sharp corners are harder to set. The edge should not be too thin, like a metal stamping or wafer-thin stone. These are awkward to fit to standard bezel wire and hard to hold onto while you wrap it.

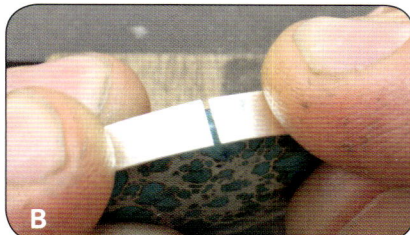

For a snug-fitting double bezel, trim the bezel wire 1mm short and later pass the stone through repeatedly until the bezel is the perfect fit.

Form and solder the bezel

For this project, I'm setting a 2" (51mm) long oval bead that is rounded on both sides. The bezel wire to be wide enough to leave enough bezel to burnish on both sides **[A]**. Be sure the bezel will cover the thickest edges of the stone with room for crimping. Wrap the bezel wire around the stone and scribe a line where the ends overlap. After soldering, the bezel should fit snugly around the stone without falling off. My trick is to trim the bezel up to 1mm short on purpose **[B]**. After soldering you can stretch the bezel by passing the stone through it repeatedly until it's the perfect fit.

Close the join with a 1mm flat chip of hard solder (see *Looking Sharp Bezel* project, p. 51). Pickle for 5 minutes.

Check the fit

This is a good time to move the join to where you want it to be hidden by the jump ring and bail. Just reposition the join and use your fingers to match the general shape. Placing the stone inside will reshape it again. Ideally, with a few passes of the stone through the bezel, it should stretch to fit tight. If the bezel is too loose, cut it open at the join and repeat the first step to adjust the fit. If it's too tight, stretch it on a ring mandrel. For large stones like this one, you many need a large-size ring mandrel (sizes 16–24) or similar to fit the bezel.

Solder on a jump ring

Center the bezel along the girdle of the stone and mark where to solder the jump ring with a permanent marker **[C]**. Soldering the ring on the bezel join will help to disguise the line. The ring

has to be horizontal, in the same plane as the bezel. If the ring is soldered on too close to the edge of the bezel, or vertical across the bezel, the solder will harden it, and then it can't be set.

Close the jump ring with a 1mm chip of hard solder. File a flat spot at the join. Flux the join on the bezel and use a third hand to hold the ring in place. Elevating the bezel on a charcoal block will make it easier to line it up. Warm the flux between the ring and bezel first until it clears. Remove the flame, and while you scoop up the solder, the flux will harden, holding the two parts together. Using a minimum of heat, place the 1mm ball of easy solder on the join. Heat from inside the bezel to draw the solder into the join **[D]**. Heating directly on the join can cause the solder to flow first to the jump ring. Pickle for 5 minutes.

Set the bezel

Setting the bezel on both sides of a stone can be slippery business, since there is no base to support the stone. I recommend gluing it around the stone first to tack it in place and keep it steady during setting. The bezel will still be

crimped firmly around the piece, capturing it securely even if the glue fails later.

Use two-part clear epoxy from the hardware store, usually sold in a double-barrel syringe. I used a fast-setting epoxy that cured in an hour. (I don't recommend E6000, since it thickens too quickly.) Mix the two parts with a toothpick on a scrap of paper. Apply a thin layer of epoxy around the inside centerline of the bezel. Insert the stone and align the bezel. Let the glue cure following the manufacturer's recommended drying time. Save the glue you mixed: When the puddle hardens, you'll know your piece is ready.

Rest the setting on leather to avoid scratching the stone. The surface underneath should be firm; use a bench pin or steel block. Crimp the bezel with a prong pusher as usual, alternating sides often. For example, crimp the first four points, and then flip the setting over and crimp the same four points on the opposite side.

Crimping puts a lot of pressure on the bezel, and working evenly on both sides will keep it from twisting around the stone.

Continue crimping in pairs of opposites **[E]** until the bezel is flush against the stone's crown on both sides. Burnish the edges of the bezel **[F]**. Burnish the bezel in opposite quadrants around the bezel, and alternate sides as you work.

Inspect the bezel closely, and push any stubborn gaps down with the prong pusher **[G]**. Burnish again. If a wrinkle is too hardened to push down, or if the bezel is loose, use flatnose pliers. Protect the back of the setting with a scrap of leather. Place the lower jaw on the leather and align the top of the pliers with the edge of the bezel where it meets the stone. The pliers have a lot of leverage, so press lightly. If the bezel is loose, pressing with the pliers can squeeze it, causing it to grip the stone. When the bezel is tight, stop! Overworking it can loosen the bezel.

Solder on a bail

Make a bail from a 6x25mm (¼x1") strip of 22-gauge sterling sheet **[H]**. Close it around the jump ring with a 1mm ball of easy solder (see *Looking Sharp Bezel* project, p. 51) **[I]**. When you work efficiently, it's possible to solder the bail on without damaging the stone. But just to be safe, use a heat shield. I placed a thick layer of nontoxic Vigor Heat Shield around the stone, leaving the bail and jump ring exposed. Let it air cool. Use a toothbrush with soap and water to remove the heat shield. Hold just the bail in hot pickle for up to a minute, protecting the stone, which could be damaged.

Polish

Polish following the recommendations for abrasive attachments below **[J]**. Using hard silicone wheels can create friction and heat, temporarily loosening the epoxy, which might loosen the stone. Let the piece cool for a few hours to allow the glue to harden again. If it's still loose, crimp and burnish again to tighten the bezel.

POLISHING ABRASIVE	WHERE TO POLISH
Silicone polishing wheel, black	The join in the bezel, excess solder, any deep scratches around the bezel, and later, the join on the bail.
Silicone polishing wheel, blue	The entire bezel, aligning the wheel parallel to the edge to avoid scratching the stone.
Polishing pin, green	Inside the jump ring and inside the bail.
Radial bristle disks: blue, peach	The jump ring, the join to the bezel, and the bail.
Radial bristle disks, light green	The entire piece, including the bail, to even out the finish.

NOUVEAU WESTERN BEZEL CUFF

This new take on the classic southwestern cuff strips the decoration to a simple twisted wire border and copper feathers. With this project, you'll practice binding parts together for soldering and learn how to heat a big cuff. It can be soldered with a large-flame butane torch, but for some of the steps, you'll need to use two of these torches at once to bring the metal up to temperature. If you have a small tank torch, the melting tip will provide enough heat.

LEVEL
Intermediate

TECHNIQUES
Sawing identical parts

Making a cuff

Using binding wire to help solder

MATERIALS
- 12" (30.5cm) 28-gauge fine silver plain bezel strip, ⅛–¼" (3–6mm) (varies with cabochon size)
- 12" (30.5cm) 6-gauge sterling round wire, dead soft
- 3x3" (76x76mm) 20-gauge copper sheet
- 6" (15.2cm) 16-gauge sterling round wire, dead soft
- 12" (30.5cm) 22-gauge sterling round wire, dead soft
- Easy, medium, and hard silver solder wire
- Cabochon
- 24- and 19-gauge dark annealed steel binding wire
- Superglue

TORCH
Two large-flame butane torches, medium to maximum flame, or small torch with melting tip, medium to large flame

FLUX
Cupronil and paste flux

tip Your cuff size is the same as your bracelet size. For example, if you wear an 8" (20.3cm) bracelet, you'd wear an 8" cuff. The difference is that a cuff has a 1–2" gap. The size of the gap determines how easily you can put the cuff over your wrist and how well it will stay on. But cuffs are dead soft after soldering, so they're malleable enough to be adjusted for a better fit.

Form the cuff

Measure twice the length for your cuff. For example, the length of a 8" (20.3cm) cuff with a 2" (51mm) gap is 6" (15.2cm); double that number to arrive at the length of wire to cut. In this example, you would saw 12" (30.5cm) of 6-gauge sterling silver round wire **[A]**. File the ends flat for a good join **[B]**. Measure the center of the wire and bend it in half over a round bezel mandrel held in a vise **[C]**. Use flat/half-round pliers to bend the ends. Make the ends parallel and overlap them a few times to create tension to close the join. If there are gaps, an easy way to fix the problem is to saw through the join with a #2 blade **[D]**. The blade will cut both ends at the same time. Hold the join closed inside a ring clamp. It may take a couple of times to saw away enough material for the ends to be flush. Be careful not to saw at an angle or you'll ruin the join and have to make a shorter cuff.

Keeping the join together for soldering can be challenging. Use dark annealed steel binding wire to make pins to hold the wire down against a soft charcoal block. Cut three or four 1" (25.5mm) pieces of 19-gauge wire and bend ¼" (6.5mm) over with roundnose pliers to make a hook.

Apply Cupronil to both sides of the cuff (see *Using spray flux*, p. 22). When it cools, press the pins around the wire, into the soft charcoal, to keep the join closed. Use the maximum flame on the butane torch or a medium flame with the melting tip to close it with 1mm balls of hard solder **[E]**. Apply additional balls one at a time until the join is full if necessary. To get the join to solder temperature, heat the whole length of the cuff until all of the flux is clear. Then concentrate the heat around the join. Pickle for 10 minutes.

To shape the end of the cuff to match the other side, first tap it with a rawhide mallet on a nylon block to round the loop **[F]**. Using nylon and rawhide will prevent marring the wire. Form it

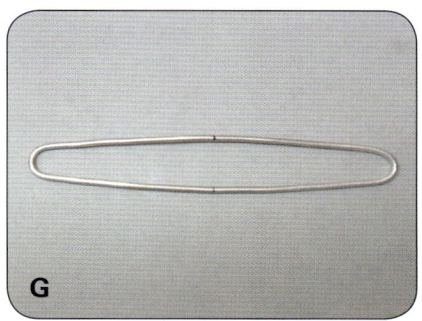

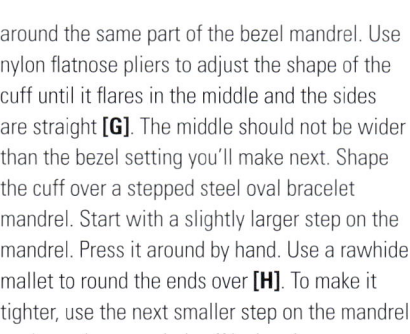

around the same part of the bezel mandrel. Use nylon flatnose pliers to adjust the shape of the cuff until it flares in the middle and the sides are straight **[G]**. The middle should not be wider than the bezel setting you'll make next. Shape the cuff over a stepped steel oval bracelet mandrel. Start with a slightly larger step on the mandrel. Press it around by hand. Use a rawhide mallet to round the ends over **[H]**. To make it tighter, use the next smaller step on the mandrel or shape the annealed cuff by hand.

Make the bezel setting

Refer to *Bezel-setting*, p. 33, and the *Looking Sharp Bezel* project, p. 51, for instructions on how to make a setting for your stone. Another way to raise the base to circulate heat from underneath it while soldering is to make a binding wire nest **[I]**. Don't be shy; use enough to make a dense wire wad. Shape it until the setting rests steady on top. Place it on a heat reflective surface, like charcoal. When the flame is aimed down into the nest, the heat will reflect back up under the base **[J]**. You can also aim the torch up and under the edge to coax the solder into flowing.

After pickling, check the fit with dental floss under the stone before continuing. Then use dividers to trace a 2–3mm border around the bezel. Saw it to size and file. Remove any excess solder with a white silicone wheel shaped to a 45-degree edge (see *Polishing With Power Tools*, p. 104). Stamp the border to make a pattern **[K]**. Support the base on a steel block and use a chasing hammer. I used a cupped 3mm nail set, available at hardware stores, to make an overlapping texture of loops. Practice first on scrap metal. File and sand the edges of the base after stamping, through 600-grit, and remove any burrs with a scraper.

To twist some wire to wrap around the bezel, cut two 6" (15.2cm) lengths of 22-gauge sterling round wire and hold the ends in a vise. Hold the other end with pliers, pull to create tension, and twist in one direction **[L]**. The twist can be tighter near one end, so remove it from the vise and clamp the opposite end. Twist again to even out the pattern. Trim one end flat with flush cutters. Use your fingers and flat/half-round pliers to fit the twisted wire around the bezel. Mark where the wire overlaps the first cut and trim flush. Try to match the join to the pattern of

twists. Solder it closed away from the setting, with Cupronil and a 1mm ball of hard solder. After pickling, it should be tight to the sides of the bezel and fit all the way down to the copper sheet. The top edge of the wire around the bezel

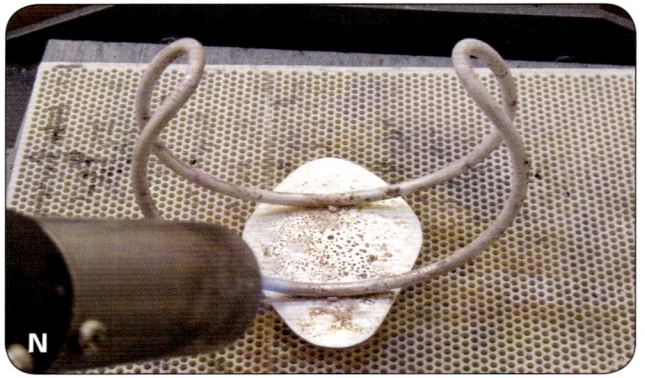

has to be below the part that will be crimped down against the stone, which is usually the top 1–2mm.

The twisted wire is soldered without getting solder into the stamped pattern. Place it around the bezel and apply Cupronil to all sides of the setting. Rest it on the wire nest on a charcoal block. Heat the setting from underneath and above with a maximum butane flame or medium melting tip flame, until the flux turns clear. Split a 1mm ball of medium silver solder in half (see *Splitting solder*, p. 29). Place a .5mm ball where the top of one twist touches the bezel. Continue heating, alternating between heating the entire setting and focusing on the join, until the solder flows **[M]**. Sometimes the solder will flow against just the top of the twist, but if the join is good, it will draw down along the wire and touch the copper sheet. Remove the heat before it flows into the stamping. It's better to place the solder at the top of the wire, because on the copper it could just as easily flow into the

stamping, which is hard to clean up. Continue working around the bezel, soldering every other loop. Pickle for 10 minutes. Inspect the joins with a loupe.

Solder the setting to the cuff

The top of the cuff will need a couple of flat spots filed to make a good join with the setting. Place the setting upside down, and balance the cuff on the back. Mark where to file the wires and transfer those marks to the front of the cuff. Brace the cuff against the bench pin and file flat spots on top of each wire with a cut 0 hand file. Test the fit. The cuff should lay flat on both spots, with the ends balanced evenly on both sides. Apply Cupronil separately to all sides of the setting and cuff. Place them on a level heat-reflective board, like honeycomb. Place a 2mm chip of medium silver solder against the first join. Use a maximum butane flame or medium oxy/propane flame to heat the entire cuff and setting. As the flux clears, dwell more

often around the join until the solder flows **[N]**. Repeat for the second join. Inspect the joins before quenching and make sure they're complete on all sides. Pickle for 10 minutes.

Make the feathers

The feathers are made of 20-gauge copper sheet and 16-gauge round sterling wire. To make matching feathers, laminate two sheets of copper so all of the sawing and filing is identical: Use shears to cut two matching rectangles of copper sheet slightly larger than the pattern, clean the metal with a Scotchbrite pad and soap, rinse, and dry thoroughly. Superglue the rectangles together, holding with pressure for a few seconds until the glue cures. Use rubber cement to glue the pattern onto the top sheet. Let it dry for 10 minutes.

Saw close to the edge of the feather outline with a #2 blade **[O]**. File away any excess beyond the pattern with a cut 0 hand file and needle files. Hold the stack in a ring clamp and refine the edges with the same files **[P]**. In a well-ventilated area, separate the pieces by heating them to about 200°F (93°C) on the solder board until you see smoke or a flame. Scrub off any discoloration with soap and water. Chip off any hardened glue using a scraper.

For texture, use the narrow wedge peen of a goldsmith's hammer to make upward angling lines on either side of the center of the feather **[Q]**. The edges can be quickly beveled, framing the feather texture, with a ball peen of a chasing hammer, striking on the edge of the sheet. File the edges again to fix any distortions from hammering. Remove file marks from the edge with a cut 0 hand file and needle files. To save polishing time later, sand the edges to a fine finish with 320- and 600-grit sanding sticks.

Worried about getting solder into the stamped pattern? Use anti-flux to protect it (see *Anti-flux*, p. 30). After applying the Cupronil, while the metal is still relatively cool, paint yellow ochre gouache over the pattern. Be sure the anti-flux doesn't touch the twisted wire where you want the solder to flow.

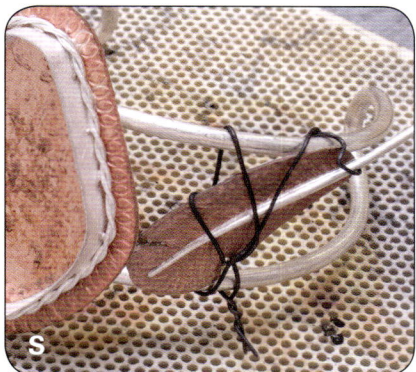
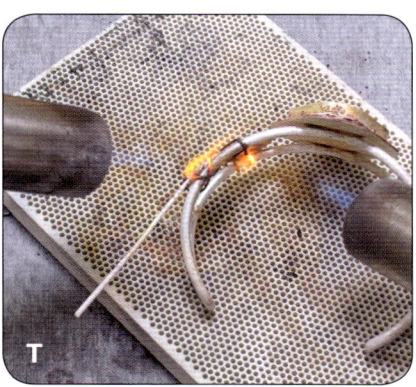
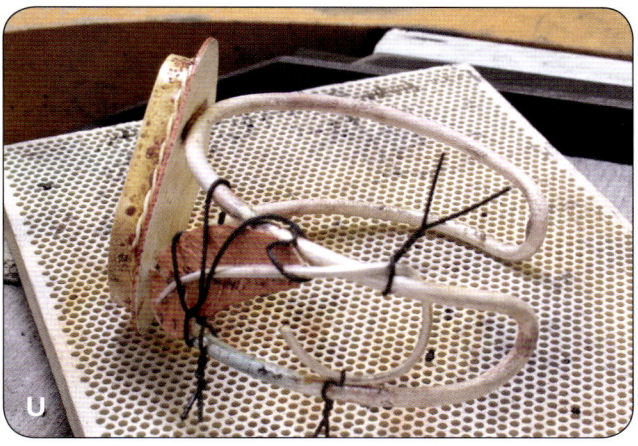

Cut two 3" (76mm) pieces of 16-gauge sterling round wire. Taper ½" (13mm) of one end on each wire (see *Mixed Metal Earrings*, p. 38). Curve the tapered end to match the curve of the feather with flat/half-round pliers. Flatten the wire with a rawhide mallet and bench block, so that it lays flat against the feather. Solder the tapered end of the wire onto each feather with one or two 1mm balls of medium silver solder **[R]**. Apply Cupronil to all sides of the feather and wire. For added protection, paint yellow ochre over the texture, but don't touch the join. Use a medium flame with a butane torch or melting tip to heat the flux on both. As it clears, the flux will help to hold the wire against the copper. Place a ball of solder near the thick end of the stem. Use your pick to hold the wire in place as you heat the copper until the solder

flows. Inspect the join line when it cools. If there is a gap, hammer it closed. Apply more Cupronil around the join and add another ball of solder. Pickle for 10 minutes.

Solder the feathers to the cuff

The feathers are soldered to the cuff at opposite angles on either side of the setting and held in place for soldering with binding wire. Curve the feather by hand or over the bracelet mandrel to match the cuff. Wrap some 24-gauge dark annealed steel binding wire around the middle of the first feather and tie it around the cuff. Twist the ends together **[S]**. The copper should rest flat against the cuff where the top and the bottom of the feather touch. Try not to touch the join itself with the steel wire; the dark scale on

the annealed wire resists solder, but it can still be soldered. Keep the sterling silver tail straight.

Solder each feather in place with a 1mm ball of easy solder inside the cuff to hide the join **[T]**. Delay pickling until both feathers and wires are soldered in place; Cupronil keeps the metal so clean that it extends your working time for multiple joins. Let the piece air cool. Bend the rest of the wire, making a nice curve so that it joins with the cuff in at least two spots. Leave the end untrimmed for fitting it to a silver ball later. Tie the wire against the cuff with binding wire **[U]**. Apply more Cupronil where it was scraped off by handling. Solder the joins with 1mm balls of easy solder. Use a large flame on the melting tip or two butane torches, if necessary, to heat the entire cuff and get the solder to flow (see next page). Let it air cool.

One chip may flow the entire distance without overflow into the texture, but it's also possible that some of the join will be made by fusing the sterling to the copper. Since sterling melts at a lower temperature than the copper, it can melt into the copper at around 1600°F (871°C), just a few hundred degrees more than the flow point of medium solder (see *Soldering Mixed Metals*, p. 32). Too hot, and the sterling will melt into a puddle. When it fuses, the wire will look like it's slightly embedded into the copper.

How to safely use two butane torches

These simple rules will help you solder and anneal safely and faster with two butane torches. Fill both torches to capacity before starting.

Use the primary torch in your nondominant hand. Light it and lock it on as usual. Apply Cupronil, bring the flux up to 1100°F (593°C), and use your other hand for picks and tweezers.

Keep the second torch on your dominant-hand side. This is your heat booster, for when you're ready to melt the solder. Practice lighting it with one hand. You should be able to release the safety and ignite it with the trigger. When used with the primary torch, the second torch isn't locked on. This makes it easy to turn it off by just letting go of the trigger. Never set it down with the flame lit; turn it off. When you set it down, make sure it's away from the soldering area.

1. **Never point a flame directly at either torch.**
2. **Use the flames to heat separate sides of the bracelet. The two flames will help overcome the conductivity of the metal that prevents one torch from being enough.**
3. **Don't cross the flames. The airflow can extinguish the other flame. If that happens, stop and reignite the torches.**
4. **Always turn off the secondary torch first, by releasing the trigger. Then unlock the first torch to turn it off.**

I made sterling balls to finish the wires. Melt two 1" (25.5mm) pieces of 16-gauge sterling wire to make two balls (see *Mixed Metal Earrings*, p. 38). Pickle them for 15 minutes. Sand off any scale around the edges.

Bend the wires to fit partially around the ball. Make a flush cut to trim it to size. The easiest way to attach a ball is to sweat-solder while holding it in place with tweezers. Since the wire around the join is thinner, one torch should be enough, with a maximum butane flame or medium flame for the melting tip. Preheat the cuff and then focus on the metal in a ½" (13mm) radius around the join. Don't melt the ends of the sterling wire!

Flux the ball and around the join with Cupronil. Melt a 1mm ball of solder on the inside of the 16-gauge wire. Hold the ball with cross-locking tweezers. Hold the ball near the flame as you heat the solder back up to its flow point. As the solder flows, press the ball against the wire **[V]**. Both the ball and the wire should be a light red. Remove the flame, but hold it in place for a few seconds to let the solder solidify. Check the join and then repeat for the second ball on the other side.

Let the cuff cool and then remove the binding wire. If there's steel in the pickle, it will plate the area around it with copper, even in citric pickle. Pickle for 20 minutes. All the soldering is done!

Add patina and polish
Darken the silver parts by applying Silver Black with a cotton swab. Paint the textured copper with hot liver of sulfur. Let the LOS sit for a couple of minutes after it turns black before rinsing in cold water. First remove the black from the highlights with a polishing pad. Use a hard pink silicone wheel to buff any stubborn areas back to a high shine.

Set the stone
Follow the usual steps to crimp and burnish the bezel around the stone. (see *Bezel-setting*, p. 33, and *Looking Sharp Bezel*, p. 51). Use the abrasive attachments recommended below to finish polishing the piece.

POLISHING ABRASIVE	WHERE TO POLISH
Silicone polishing wheel, black	Any visible join lines or excess solder. Remove scratches on the bezel or cuff, but avoid any texture.
Silicone polishing wheel, blue	The entire bezel.
Radial bristle disks: red, blue, peach, light green	The entire cuff. The flexible radials can get into the texture and twisted wire, and reach around the cuff wires to polish the back of the setting and feathers.
Polishing pins: gray, brown or green	Any hard-to-reach places, like between the feather and wires.

PEDESTAL PRONG PENDANT

Prongs adapt easily to setting unusual stones, like these rough-top tourmaline cookies. The same technique could be used with cabochons or, with a little modification, faceted gemstones with pavilions. And I love the look of the horn-shaped bail with the chunky square prongs.

LEVEL
Intermediate

TECHNIQUES
Fabricating a prong setting

Using soft charcoal to position parts for soldering

Prong-setting

MATERIALS
- 20mm cabochon or flat-bottom stone with straight sides

- 6–12" (15.2–30.5cm) 14-gauge copper wire, square, dead soft (length varies with stone)

- 6" (15.2cm) 16-gauge sterling wire, square, dead soft

- 8" (20.3cm) 18-gauge sterling wire, round, dead soft

- 2" (51mm) ³⁄₃₂" (2.4mm) OD copper seamless tubing

- Easy and hard silver solder wire

TORCH
Micro butane torch, maximum flame, or small torch with #5 or 7 tip, medium flame

FLUX
Cupronil

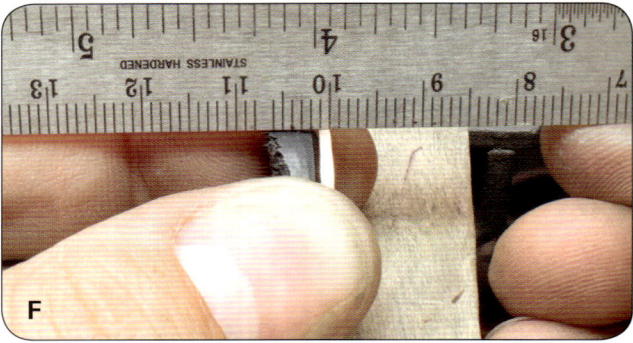

Stones for prong setting

The tourmaline I used for this setting is 14x19mm. As you choose a stone, try to picture the prongs being pushed over it. The ideal stone cut has consistent straight sides or it rounds inward like a cabochon. If the sides flare out or curve in randomly, it will be much harder to set. If a faceted stone with a pavilion is used, the pedestal will have to be raised high enough to accommodate the pointy culet. And the pedestal will need to be beveled to the same angle of the pavilion. (See *Prongs*, p. 34, for more information.)

Form the pedestal

The pedestal supports the stone and the prongs. It acts as the base for the prongs to press the stone against, making a secure setting. Saw and file one end of 14-gauge square copper wire to a flat 90-degree angle. Use flat/half-round pliers to bend it to match the size and shape of the bottom of the stone [A]. Saw it to size with a #2 blade [B]. Close the join and double-check the fit. If there are gaps, saw through the join (see *Nouveau Western Bezel Cuff*, p. 59) The outside of the pedestal should be the same size as the stone so the prongs fit flush against the girdle [C].

Set up to solder on charcoal or honeycomb. Apply Cupronil to all sides of the pedestal (see *Using spray flux*, p. 22). Solder the join closed with a 1mm ball of hard solder [D]. Pickle for 5 minutes.

After pickling, check the pedestal against the stone. If it's too small, forge it gently with the flat face of a goldsmith's hammer on a steel block to enlarge it. Try to hit flat on the square sides of the copper to avoid hammer marks. Check the fit often. If it's a little out of shape and too hard to bend with half-round pliers, stand it on edge and use the same hammer to tap it into place [E]. Once it fits, sand the top and bottom flat on 320- and 600-grit sanding sticks.

Solder on the prongs

The prongs start out longer than necessary so they can be embedded in charcoal later for soldering. Measure the length from the bottom of the pedestal to the tallest side of the stone [F]. In this example, the length is 7mm. Double that to find the length of 16-gauge sterling square wire to cut for each of the prongs. I used five prongs to set my stone. Straighten each

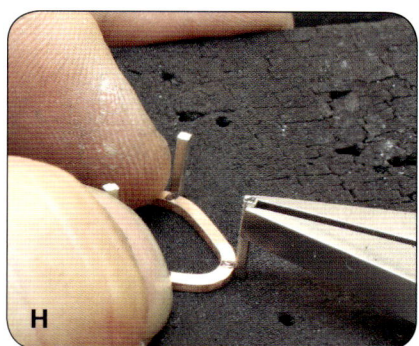

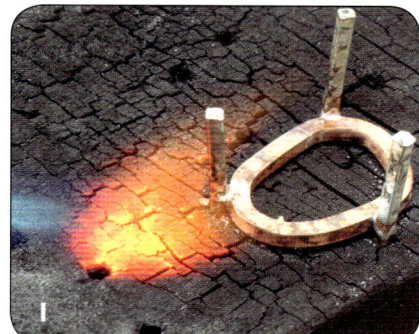

prong with a rawhide mallet and steel block. Mark the locations for the prongs with a permanent marker **[G]**. The prongs should be in opposing pairs as much as possible.

Place the unfluxed pedestal on flat, soft charcoal and press several prongs in place with pliers, embedding them almost halfway into the charcoal **[H]**. The prongs should be flush against the pedestal, holding it in place, and be straight and perpendicular. Since the marker will fade during soldering, scribe clear lines in the charcoal for the locations of the remaining prongs for later.

Leave the parts in place and apply Cupronil. The base and embedded wires will be kept clean by the charcoal because they can't move during soldering. Solder each prong in place with one or two 1mm balls of hard silver solder, placed either on the top or lower corners **[I]**. Inspect the joins to make sure they're complete on both sides.

Let the piece air cool and then embed the remaining prongs. Apply more Cupronil, and solder the prongs. While the piece is still warm, insert your tweezers and pick under the pedestal to gently pry it up a little at a time. Try not to warp the soft metal. Pickle for 10 minutes.

Trim the prongs flush with the bottom of the pedestal with flush cutters. Mark each prong at 3mm (⅛") above the top edge of the stone. Trim with flush cutters **[J]**. It's better to make the prongs slightly longer, because they can be trimmed during setting. But if they're too stubby, they're much harder to push down. For this setting, I used a file to bevel the tip of each prong at 45 degrees.

Make the bail

The tubing is bent with nylon jaw bracelet-forming pliers **[K]**. Measure 1" (25.5mm) across the curve and saw it to size with a #2 blade. Hold the tubing with a ring clamp, and support it on the bench pin to keep it from collapsing during sawing. Saw across the tube, like a hacksaw. If the blade gets stuck, use more wax and cut on the smooth forward stroke until it saws easily again **[L]**. File and sand the ends with sticks through 600-grit.

For a little added detail, I added a coil of jump rings to the bail. Coil some 18-gauge round sterling silver wire around the same diameter tubing, making about six rings. Clamp the tube against the bench pin and hold the coil with your fingers or a smaller clamp. Slide it to the end of the tube and saw at a 45-degree angle with a 2/0 blade **[M]**. As you saw through the rings,

they will drop into your sweeps drawer with parallel cut joins.

Close the rings with hard solder. Cutting the join makes them a little bit too small, but they can be stretched. Even slender bezel mandrels are usually too big to fit inside these rings, but the jaws of roundnose pliers are like tiny mandrels. Tap the ring down one of the jaws, stretching it with a nylon or rawhide mallet. It only takes a little stretching, so check the fit often until the ring fits tight. Slide all the rings to the center and align all the joins on the side that will be soldered to the setting. If the rings are too loose, they can move out of center and are difficult to resolder.

Apply Cupronil and solder between each ring with 1mm balls of hard solder **[N]**. Pickle for 10 minutes. Make sure the solder flowed through to the copper and that none of the rings still move. Hold the bail with a ring clamp and make a flat spot along the rings with #2 flat file **[O]**. Flux all sides of the bail and setting with Cupronil. The bail is thick, so to center the flat spot against the pedestal, press it down into the soft charcoal. If it's hard to press, dig a little bit of the charcoal out and try again. Reapply Cupronil to cover any bare areas. Place a 2mm

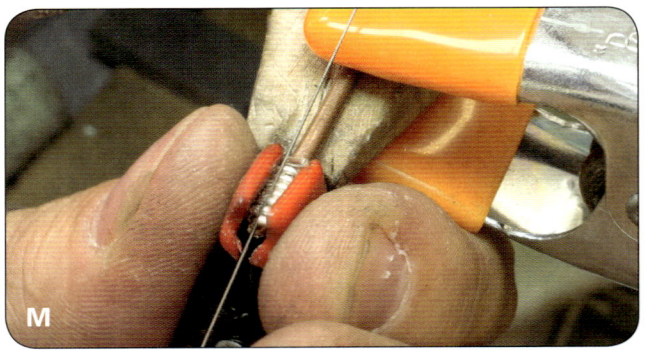

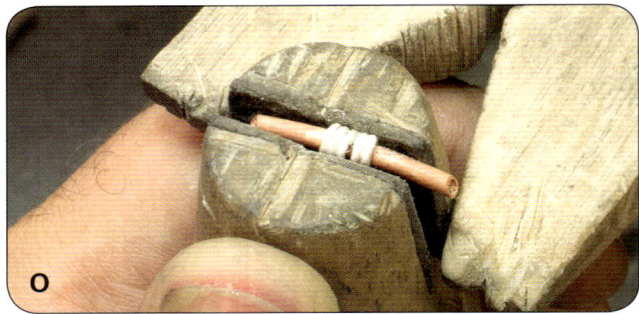

chip of easy solder on top of the join and solder **[P]**. Do not overheat the prongs or they can slip out of alignment. Pickle for 10 minutes.

Modify the prongs

Prongs need to be cut, almost like a notch into a tree, to make them easier to push down around the stone. For this rough top stone, I'm using a 3mm ball bur to thin the tops. First, with the stone in place, mark the depth and length of the cuts on the sides of the prongs **[Q]**. The cuts should be no more than halfway through the wire.

Burs have a lot of traction and can easily slip and mar your metal. Rest one finger against the smooth shaft as you work, to provide resistance. Brace the setting and your hand on the bench pin, with a scrap of leather to protect the finish. Turn the prong as you work so that the bur is always pulled toward you, not over the far edge, which will draw the bur around the prong and damage it. Use low speed and pressure to cut. The bur is harder to control at high speed. Use good light and close magnification. Dip the bur in bur lube, and draw the bur along the side you can see, making repeated cuts down close to the final depth **[R]**. Turn the prong and cut toward the top of the prong, along the other edge **[S]**. Working at a slight angle helps with control. Use the bur like a rotating tiny file to make an even cut. Repeat for each prong, checking your work against the stone. Notice that the tops of the prongs are left at their original thickness, so that they don't look thin and fragile when set **[T]**.

Polish and add patina

Polish the setting using the abrasive attachments recommended below before setting the stone. Paint hot liver of sulfur with a brush

POLISHING ABRASIVE	WHERE TO POLISH
Mini half-round file, #2 cut	Any excess lumps of solder and the join line.
Silicone polishing wheels: black, blue	Grind the edge of the wheels to a 45-degree angle against a separating disk. Polish the flat sides of the prongs and any accessible parts of the pedestal. The blue wheels should leave a high polish on the prongs. Don't polish the tops of the prongs until after setting.
Polishing pins: gray, brown, green	Inside the pedestal and any hard-to-reach areas. Polish to a mirror finish.
Radial bristle disks: blue, peach, light green	Blue radials on the bail to remove any light scale, and the inside of the prongs. Continue polishing the entire piece with peach and green.
Burnishers	Use steel burnishers to polish any hard-to-reach areas. Blend any streaks with light green radial bristle disks.

directly on the sterling rings, antiquing the lines between them and where they meet the copper tube and bail. Leave the rest of the setting without patina. Bring back the highlights with a polish pad and/or pink silicone wheel. Buff it back to a higher luster with another round of polishing with light-green radial bristle disks.

Set the stone

Rest the setting on a steel block or bench pin, and protect the finish with leather. A firm work surface helps with the pressure to set the prongs. Press the prongs against the sides of the stone with a pusher [U]. Place it on the cut section of a prong and push it over until it nearly touches the stone [V]. Push down opposing pairs of prongs, one at a time, just short of the stone. This will hold and center the stone in the setting, and leave room for final adjustments.

The prongs should extend 2–3mm over the edge of the stone [W]. Trim back any long ones with flush cutters. Use a fine blue silicone wheel with 45-degree angle on its edge to round each prong tip [X]. This will remove any sharp burs and corners. Brush away any dust. The soft silicone is friendly to most stones, but avoid hitting the stone anyway, just to be safe.

Notice how, after pushing the prongs down the rest of the way, they should be thick with only slightly beveled ends [Y]. This makes them look strong and well crafted. Any prongs that are hard to press down can be closed with parallel-jaw or chainnose pliers [Z].

Protect the pedestal with leather and don't use a lot of pressure; the pliers have a lot of leverage. Polish any work marks from setting with blue silicone wheels and light green radial bristle disks.

HOLLOW BIRD AND BRANCH PENDANT

Part of the design of this hollow bird pendant is made by mixing metals. Sterling silver for the eye and sides contrasts beautifully with the copper, especially as it tarnishes to a deeper color. The hollow form makes the piece comfortably light for wearing and is a good technique to use for earrings too.

LEVEL
Intermediate

TECHNIQUES
Soldering hollow forms

Using anti-flux

MATERIALS
- 12" (30.5cm) 24-gauge ³⁄₁₆" (5mm) sterling silver strip
- 3x3" (76x76mm) 20-gauge copper sheet
- **2** 2mm ID 20-gauge sterling silver jump rings
- ³⁄₁₆" (5mm) 2.5mm OD sterling tubing
- 2" (51mm) 2.5mm OD copper tubing
- 3" (76mm) 16-gauge sterling round wire, dead soft
- 3" (76mm) 18-gauge sterling round wire, dead soft
- 2x2" (51x51mm) scrap of ⅛" (3mm) thick balsa wood

TORCH
Large-flame butane torch, maximum flame or small torch with #5 tip and a melting tip, medium flame

FLUX
Cupronil and paste flux

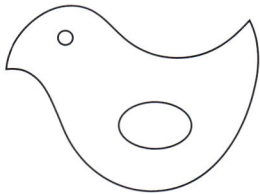

template, actual size

Make a wood template

File and saw a balsa wood version of the template above to help you form the walls of the hollow bird box.

Form the sterling strip

Cut the end of the sterling silver strip flush at 90 degrees, as you would for a bezel. Start forming the end of the wire at the point of the beak. Use flat/half-round pliers to form the curves and fit them against the balsa wood template **[A]**. Mark where to bend the tail. To make a crisp bend, use flatnose pliers. Curve the strip around the bottom of the template and up to the beak. Scribe a line where it overlaps and cut it flush **[B]**. Adjust the tension in the strip, back and forth, until the join stays closed. File it with a flat needle file, if necessary, to make a flush join.

Apply Cupronil to the entire bird form. Forge the end of some hard silver solder wire flat. Place a 1mm flat chip under the join on a flat solder surface like honeycomb. Heating from above will draw the solder from underneath, through to the top, to fill the join **[C]**. Pickle for 10 minutes.

To make the oval for the opening, start again with a flush-cut end. Use flat/half-round pliers to bend the strip into an oval that is close to the size of the template. After soldering it with hard solder and pickling, use an oval bezel mandrel and rawhide mallet to refine the shape. Sand the top and bottom edges of both pieces on 320-grit paper until they lie completely flat on the steel block with no gaps. Reshape the bird if necessary before adding the base.

Solder on the base

Cut two pieces of 20-gauge copper sheet that are 3mm (⅛") larger than the bird on all sides. Texture one side of each sheet. Flatten them with a rawhide mallet and steel block. If the sheet is just barely curved, you may be able to sand the solder side flat on 320-grit paper. If not, anneal it using Cupronil (see *Annealing*, p. 101)

and flatten it again after pickling. After you have a good join between the pieces, apply Cupronil to the sheet, bird, and oval completely. Set the bird on the copper sheet, texture side down, on a soldering tripod, as you would a bezel (see *Looking Sharp Bezel*, p. 51). Use your pick to stroke the molten solder across any sluggish spots. Add the oval shape with two or three 1mm chips of medium solder, and solder it to the base **[D]**. Pickle for 15 minutes.

Center-punch a divot inside the oval on a steel block. Drill a small hole. Pierce and saw away the excess copper inside the oval. Saw off the excess sheet around the bird, too. File close to the join, but don't remove the join line: If you file away extra solder or evidence of the join at this point, it will flow again during the next soldering, and you'll have to do it again!

Solder the top

Before soldering on the top sheet, drill a hole for the eye. The hole should match the OD of the tubing exactly for a tight solder join. Drill a pilot hole first with a drill bit that is ¹⁄₁₆" (1.6mm) or smaller. Expand it close to the right size with a ³⁄₃₂" drill bit. Either file it to fit with a round needle file, or use a 2.5mm bud or ball bur.

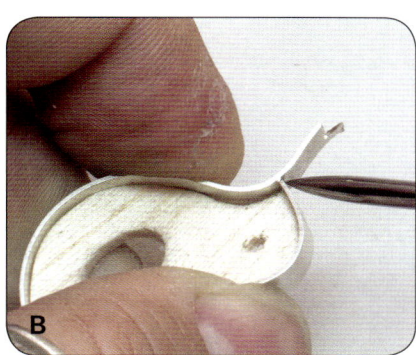

For the best joins for soldering, sand the top edges of the bird and oval. A good trick for checking to see if they are flat is to stain the edges with a permanent marker and sand it on a sheet of 320-grit sandpaper. The sandpaper will leave marker ink anywhere the walls aren't completely flat. Flatten the copper sheet and test the join.

Apply Cupronil and solder the blank side of the copper sheet to the bird with medium solder. Heat first from underneath until the flux turns clear. The mass of the box will require you to bring the flame to the top, aiming at a 45-degree angle around the join as you slowly turn the mesh to get the solder to flow **[E]**. Pickle for 10 minutes.

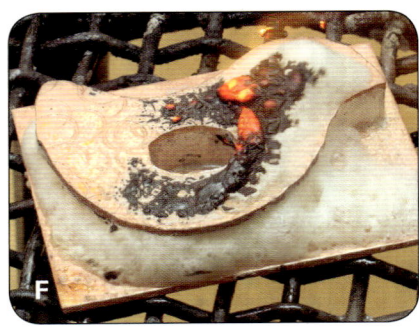

Since the shape has holes, water, pickle, and flux will be trapped inside. The next time it's soldered, a gooey mess will bubble out, which can stop the solder from working **[F]**. Before drilling or filing, remove the water by lightly heating it without any flux. Simmer the heat on and off, until the water starts steaming. The metal may turn a golden straw color. Don't heat past that point. Water will come out of even the smallest gap, so this is a good test of your soldering skills. Let the piece air cool and scrub it with a damp brass brush. Wipe it with a damp towel but don't rinse it or you'll fill it with water again.

The second hole for the tubing is best drilled with a drill press, which will keep the drill bit at 90 degrees. Use the first hole to line up the drill, placing it inside the hole while it's turned off. Drill at low speed. Avoid wiggling the box, which can enlarge the holes. Open them to the right size with a 2.5mm bud or ball bur, or with a round needle file. Check the fit with the tubing and mark where to cut it (about 1mm past the copper on both sides). If the tubing is snug inside the holes, leave it inside and use a #2 saw blade to trim it. It's harder to insert a short length of tubing than to saw it while it's in place. If there are gaps in the join, the tubing can be flared to fit with a center punch. Rest one end on a steel block, and tap the punch into the tubing with a chasing hammer.

safety

After this next soldering, the box will be sealed, trapping air inside. A jump ring still has to be soldered on, but if there's no vent, pressure will build and the joins can pop open (sometimes with a lot of force). Before soldering the tubing in place, make a divot and drill a small 1mm hole on the bottom or back where it will be the least obvious. Since a center punch will dent the soft metal, use a 1mm ball bur to make a divot to help align the drill bit.

It's very easy to overflow the solder into the texture on the copper while soldering the sterling silver tubing into the shape. Use anti-flux, like yellow ochre gouache, which will dirty the copper and prevent the solder from flowing (see *Anti-flux*, p. 30).

Place a 1mm ball of easy solder on the join and heat from above and below. The yellow ochre will burn to a black crust. Flip the piece over and let it cool for a minute. Apply a coat of yellow ochre around the second join before soldering. Let it air cool; quenching will get water inside again. Scrub off the yellow ochre with a Scotchbrite pad. Before pickling, plug the vent hole with a toothpick, trimming it with wire cutters. If the pick is a tight fit, no pickle will get inside, saving you time.

Solder the jump ring

After pickling, remove the toothpick, and remove any water that got inside. Polish the top of the bird through 600-grit where the jump ring will be attached. File a small flat spot at the join. Before soldering the jump ring to the bird, think about how to balance it: The ring may need to be a little off-center so the shape hangs straight.

Solder a 2mm 20-gauge sterling jump ring with hard solder: Set up to solder on a honeycomb board. Apply Cupronil flux to the bird. Hold the ring in place with a third hand. Melt a 1mm ball of easy solder into the join **[G]**. Most of the heat needs to be on the bird itself first, moving to the join as the solder starts to flow. Let the piece air cool.

After soldering, insert your pick into the ring and check the balance. Unsolder and move it if necessary (see *Unsoldering*, p. 30).

Cork the hole with a new toothpick. Pickle for 10 minutes. Polish the rest of the bird before attaching the bail.

Form the branch bail

The branch bail with wire leaves and vines accents the bird theme while providing a connection for a chain or beadwork. Tubing will crimp if it's bent too far, but a shallow curve can be made with nylon bracelet-forming pliers. Saw the tubing to 1" (25.5mm) and file the ends with a cut 0 hand file. Finish them with 320- and 600-grit sanding sticks. Form the tubing into a gentle curve.

For the first vine, cut 2" (51mm) of 16-gauge sterling silver round wire. With the polished face of a goldsmith's hammer, forge one end into a flat wedge to make a leaf. Do not make it too thin. File it into a pointed leaf shape with a 2-cut needle file. Sand the edges and leaf to a 600-grit finish before soldering. It's easier to wrap the wire if one end is soldered to the tubing first: Apply Cupronil to the tubing and wire and set up on a honeycomb board. Rest the wire in place and solder it with a 1mm ball of hard silver solder **[H]**. Pickle for 2 minutes to remove the flux.

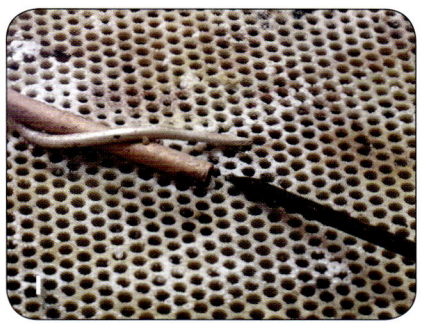

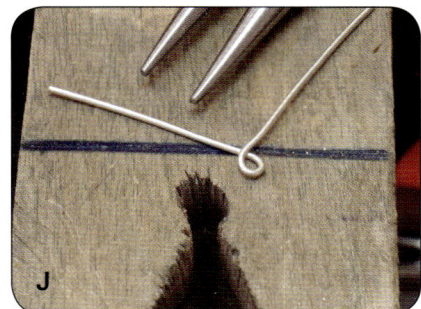

Wrap the wire tightly around the tubing, but be careful not to crush it. Forge the end into a leaf shape, but wait to wrap it until it anneals during soldering. Apply Cupronil again and use the same solder to join the wire to the tubing in two places: where it touches at the other end and somewhere in the middle **[I]**. Quench and dry. Wrap the leaf around the tubing. Apply more Cupronil and solder it. Pickle for 2 minutes.

The second vine is made of 18-gauge sterling silver wire for contrast, and will also form the loop to connect it to the bird. The wire ends can be balled and flattened or forged and filed into small leaf shapes.

Trim a piece of wire to 3" (76mm) and cross the ends to make a loop in the center **[J]**. Coil each end around the tubing, mixing it with the 16-gauge wire and leaving ½" (13mm) of each end raised. Keep the coil centered and in line with the tubing. Tack the wires in place with hard solder. Forge the ends of the wires lightly and file into smaller leaf shapes, if desired.

Bend them against the tubing with nylon flatnose pliers. Make sure there is a good join on either end between the new wire and the bail to secure it **[K]**. Apply Cupronil again and solder both ends with 1mm balls of hard solder. Pickle for 10 minutes. Inspect the joins and polish the bail before connecting it to the bird.

Hide the hole
The small vent hole can be plugged with some matching wire: File a taper on one end of a piece of 16-gauge wire. Screw it tightly into the hole. Cut it flush to the surface and use a black silicone wheel to polish it down to match the metal. It won't be invisible, but it disguises the hole beautifully.

Polish and add patina
It's easiest to polish the bird and bail separately before connecting them with the jump ring. Polish using the abrasive attachments in steps recommended below. After polishing, paint hot liver of sulfur with a brush directly on the copper texture, on the front and back, inside the bail,

and where the tubing needs to be darkened around the wires. Let it sit for a few minutes to turn black. Bring back the highlights with a polishing pad and/or pink silicone wheel. Blend away any streaks with another round of light green radial bristle disks.

The bail is connected with a 2mm 20-gauge sterling jump ring: Open the ring, insert the loops from the bird and bail, and close the ring for soldering.

Hold the bail loop or the jump ring on the bird (not the jump ring to be soldered) with a third hand (see *Dog Tag Chain*, p. 42). Coat the jump rings, bail, and bird with paste flux, which is easiest to apply for this kind of join. Split a 1mm ball of solder into a .5mm ball and use a small butane torch flame or a #5 tip with a medium flame to solder the jump ring closed.

Pickle only the jump rings (for 2 minutes or less). Cork the hole in the bird, just in case it falls in.

POLISHING ABRASIVE	WHERE TO POLISH
Stack of 4 separating disks and/or files	Remove the rest of the excess copper around the joins on the outside. Stop when the join is almost gone to avoid scratching the sterling. Use half-round needle files to clean up the joins inside the oval. Grind down the excess tubing too.
Silicone polishing wheel, white	Remove any scratches from the separating disks and finish the joins. Avoid making divots or waves in the surface.
Split mandrels: 320-, 600-,1200-grit	Sand the top, bottom, and sides of the bird, and inside the oval, through 1200-grit. Stop just short of the jump ring.
Silicone polishing wheels: black, blue	Carefully polish near the jump ring to blend it in with the rest of the box.
Radial bristle disks: blue 400-grit	The bail. After connecting the jump ring, polish it too.
Radial bristle disks: peach, light green	The bail and bird to a mirror finish. The radial bristle disks won't fit inside the oval. Polish the jump rings after soldering.
Silicone polishing cylinders: blue, pink	Inside the oval with blue and pink. Taper the cylinders to a point on a separating disk. Use the tip of the pink cylinder to polish inside any rings.

CROWN PRONG RINGS

As always, something simple and classic looking, such as a solitaire four-prong ring, hides a lot of skill and a few tricks. This project lays out step-by-step tips for notching and setting prongs around a faceted gemstone to make a stand-up-and-shout crown ring. After you make a single-stone ring, explore variations like the one shown above left.

LEVEL
Intermediate

TECHNIQUES
Prong-setting with setting burs

Making a ring

MATERIALS
6" (15.2cm) 10-gauge sterling half-round wire, dead soft

Sterling 8mm round low-base 4-prong setting

6mm ID 18-gauge sterling round jump ring

Easy, medium, and hard silver solder

8mm round faceted gemstone

2 sterling 4mm round low base 4-prong settings (optional)

2 3.2mm ID 18-gauge sterling round jump rings (optional)

2 4mm round faceted gemstones (optional)

TORCH
Micro butane torch, maximum flame, or small torch with #5 or #7 tip, medium flame

FLUX
Cupronil

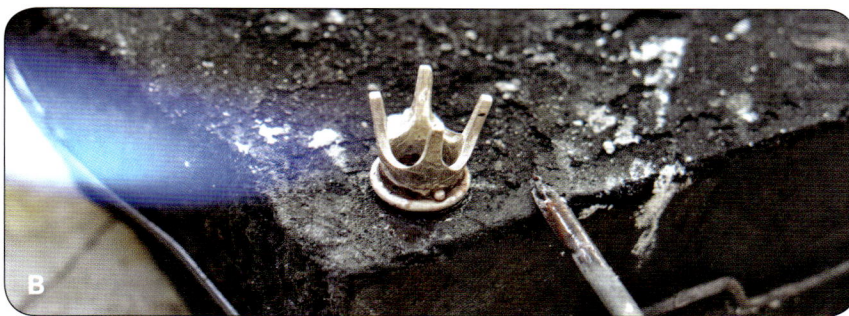

Choose a stone and setting

Prong settings are available in sterling and gold from jewelry suppliers. They are listed by the size of stone they fit. Since prongs are flexible, they can be adjusted to custom fit a slightly smaller or larger stone. When you rest your stone on top of the right size finding, the girdle will hit the middle of the top of each prong.

Your first stone should be easy to set, like a round machine-cut stone, with a well-cut crown and pavilion. After a few successful settings, you'll be ready to tackle stones with unusual cuts, like natural gemstones, which can have very rounded pavilions that require customizing the prongs with ball burs. For more information, see *Stone Setting Techniques: Anatomy of a stone*, p. 33, and *Prongs*, p. 34.

Make a ring

Making a ring is a beginning jewelry skill, but we'll review some of it here. Use finger gauges or measure a ring that fits on a ring mandrel. Find the blank length for your ring size on the Ring Blank Size by Gauge chart, p. 108. The gauge of half-round wire refers to the widest part, not the thickness, which is what you want to reference on the size chart.

For example, 10-gauge half-round is 2.6mm wide and 1.3mm thick. Cross-reference the 1.3mm (16-gauge) column on the chart with your size to find the length to cut. Lock that measurement on digital calipers. Cut one end of the 10-gauge half-round wire with heavy-duty flush cutters. Place one jaw of the calipers on the flat end and scribe a line with the sharp point of the other jaw. Cut the other side flush.

Both ends should be flat and 90 degrees. File if necessary.

To form the ring, start by curving the ends. Lay the blank across a ring mandrel at the correct size, with ½" (13mm) extended past the steel. Use a rawhide mallet to curve the end against the mandrel **[A]**. Repeat on the other end of the blank. Continue extending and hammering the blank against the mandrel to form a round ring. The ends won't touch, but they can be aligned and closed with flat/half-round pliers. File or saw the join to remove any gaps for a flush fit. Set up to solder on a charcoal block or honeycomb board. Apply Cupronil and close the ring with a 1mm ball of hard solder. Pickle for 5 minutes. Reshape the ring with the mandrel and rawhide mallet. Check the size when the ring is tight against the steel with no gaps (see *Adjusting Ring Size*, p. 108).

Attach the setting

The base of the setting is accented with a wire border made either from 18-gauge sterling round wire or a matching jump ring. Solder a 6mm ID 18-gauge sterling jump ring closed with a 1mm ball of hard solder. Pickle for 2 minutes to remove the flux. The base of the setting should sit flush to the bottom of the ring with no gaps for a good solder join. If it's too small, the setting will tilt. Stretch it on a round bezel mandrel, tapping it down with the corner of a rawhide mallet. Check the fit often, before it gets too large. Apply Cupronil and join the ends with a 1mm ball of hard solder **[B]**. Pickle for 2 minutes and inspect the join before attaching it to the ring.

It's easiest to balance and solder the setting onto the ring if a flat spot is filed first. Use a cut 0 hand file and file it at the join, to the width of the base of the setting, but not wider **[C]**. Set the crown prong-side down on the charcoal block. Apply Cupronil and hold the ring in place with a third hand. One pair of prongs should line up with the center of the ring so it looks symmetrical. Place a 1mm ball of medium solder on either side of the join **[D]**. Continue heating, balancing the heat between the ring and the setting until the solder is drawn into both joins. Pickle for 10 minutes.

The prongs on cast settings are fragile during soldering. Too much pressure can crack them, especially while they are red-hot. Make adjustments during soldering by handling the setting by the base.

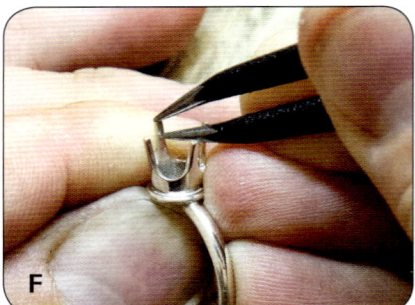

Multiple settings

A variation of this ring adds two 4mm prong settings. After soldering on the first, pickle it. File flat spots for each of the new settings. Apply Cupronil again and use the third hand to line up each one for soldering, one at a time **[E]**. Be careful not to overheat the other two settings or they can move. As a precaution, you can paint the joins on the other settings with anti-flux.

Polish and add patina

Polish the ring before setting using the abrasive attachments recommended below. A mirror polish inside the prong setting will accent the color of the stone by reflecting more light.

The border around the setting looks nice with a little antiquing. Use a cotton swab to paint Silver Black along the border with the setting and where it meets the ring, and remove the excess with a radial bristle disk.

Set the stone

A properly made prong setting has a level seat for the stone and the prongs are squared with the facets. The prongs look thick and have slight, rounded bevels. This is delicate work, so use magnification and good lighting to see what you're doing. Steady the bur by placing one finger on the smooth, slowly spinning shaft for added control.

Measure the height from the table to the girdle with dividers and transfer it to the inside of each prong. Scribe the line over to the sides of the prong for a clear reference **[F]**. This line is where the wide girdle will sit. With a 5mm setting bur, which is not so large that it will accidentally cut the other prongs as you work, notch each of the prongs at the mark **[G]**. Lubricate the bur with bur lube. Tilt the bur slightly, so that the corner of the bur is at the scribed line. With low speed and pressure, grind almost halfway through

the prong. The top of the prong should be the original thickness, and there should be two clear angles cut straight across the silver **[H]**. Work the side you can see; working on the far edge can catch the bur and damage the prong. Repeat this cut on each prong. As you rotate the setting, even out the cut by moving the bur, working it against the top edge **[I]**. In this position, the bur is cutting the other side of the prong, and it is easier to see the profile as you work.

Insert the stone by resting the table on the bench pin and pressing the setting down on top. If the seat is cut well, the stone will snap in and is easier to level **[J]** . If this doesn't work, use tweezers or some sticky wax on a toothpick to drop it into the setting from above. Brush off any dust and inspect each prong with a loupe. Check the fit. The pavilion should rest on the lower cut of each prong and the table should be level with the top of the setting **[K]**.

POLISHING ABRASIVE	WHERE TO POLISH
Silicone polishing wheels: black	The joins on the ring and border around the setting. Use a smaller, worn wheel to work inside the ring. Remove any excess solder. Remove any hammer marks from stretching the ring.
Split mandrel: 320-, 600-, 1200-grit	Sand the inside of the ring to a satin shine.
Polishing pins: gray, brown, green (or silicone cylinders: blue, pink)	Inside the prong setting. For larger settings, taper silicone cylinders against a separation disk. Don't thin the prongs!
Radial bristle disks: blue, peach, light green	Use the blue disks to remove any light scale on the outside of the ring and setting. Continue polishing the entire piece with peach and green, including inside the ring. Don't round the sharp corners on the prongs.

J

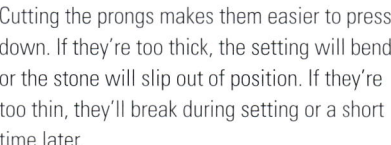

K

L

Cutting the prongs makes them easier to press down. If they're too thick, the setting will bend or the stone will slip out of position. If they're too thin, they'll break during setting or a short time later.

Hold the ring in a ring clamp. Rotate the stone before setting, so that the square facets in the table are aligned with the prongs. Tighten the prongs against the stone with smooth chainnose pliers, squeezing gently just below the girdle **[L]**. The next adjustment is odd, but it tightens the prongs further, making the next step easier. If the prongs were numbered 1–4, hold number 1 and 2 below the girdle with chainnose and squeeze lightly. The prongs will barely move, but they will tighten against the stone. Repeat for the opposite pair, 3 and 4. This will move the prongs out of square, so realign them to the facets by tightening the other pairs the same way: 1 and 4, 2 and 3 **[M]**.

Time to lower the prongs: Lower each prong only three-quarters of the way down in the first round. Pressing it all the way can tilt the stone and doesn't leave space to safely shape the prongs. Work in pairs of opposites around the setting. Rest one jaw of the chainnose against

M

the side of a prong and press the opposite prong down most of the way **[N]**. Reverse the hold and press the opposite prong down. Repeat for the second pair. Check to make sure the stone is still level.

The ends of the prongs, when set completely, should touch the corners of the square facets on the table. If they're too long, they can be trimmed with flush cutters. To finish setting the prongs, move one jaw to the wire border, for more leverage. Lower each prong, in pairs of opposites, down to the crown **[O]**. Check your work with a loupe and correct any gaps. Usually there is a slight gap at the end of the prong. When the end is rounded with a cup bur, it creates a lip of metal that can fill in that gap **[P]**. The cup bur should be large enough to fit around the front without grinding into the back of the prong. For the 8mm setting, I used a 2mm cup bur. Hold the bur at a forward angle with the flex shaft at low speed, so that the teeth round the front edge without flattening the prong. Bring back the high shine with a blue silicone wheel, shaped to a 45-degree angle. Polish it again with light green for a bright finish.

N

O

Troubleshooting

Too much pressure and the stone can tilt or the setting can bend. If the stone tilts, gently pry back the prongs with a prong lifter just enough to level the stone and then press the prongs down again. If it continues to tilt, the seat may have been burred at an angle or not deeply enough to hold the stone against the pressure of the prongs. If that's the case, you may have to remove the stone and bur the seats again for a better fit. If the setting bends, remove the stone and use the pliers to readjust the setting. Too much back-and-forth movement on the prongs can break them, so be gentle. The setting can be annealed to soften them again. Apply Cupronil to the entire ring to preserve the finish. Paint the joins with anti-flux and hold the ring with a third hand to balance the settings so that they don't move! Heat it until the flux turns clear, and the silver glows a light pink. Quench, pickle, and restore the finish.

P

RECYCLED STERLING RING

Make this ring from scraps of sterling silver sheet. I'm demonstrating two styles of fusing in this project: textured and nontextured. The easiest way to fuse is to allow the metals to go a little beyond fusion, melting them together and creating reticulated textures. Once you've observed the color and gloss of the surface as it fuses, it's easier to recognize the first stages and pull away, preserving the original texture of the metal.

Use this technique to recycle scrap into big, textured sheets. With a melting tip on a small torch, you can make bracelet-size blanks!

LEVEL
Beginner

TECHNIQUES
Fusing sterling silver

Fusing Argentium
sterling silver

MATERIALS
- 1x3" (25.5x76mm) 22- or 24-gauge sterling silver sheet
- 12" (30.5cm) Argentium sterling silver 14-gauge round wire, dead soft
- Hard silver solder

TORCH
Large-flame butane torch, maximum flame or small torch with melting tip, medium flame

FLUX
Paste flux and Cupronil

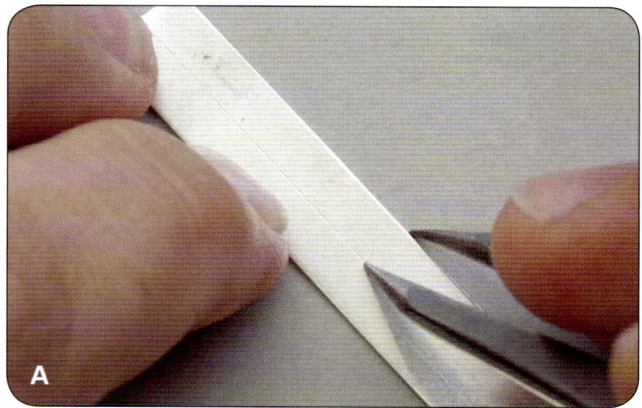

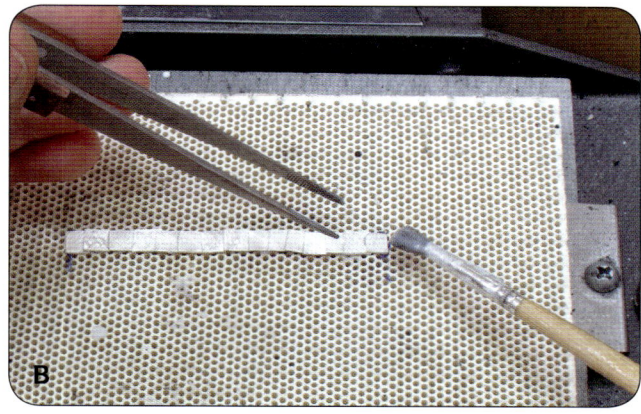

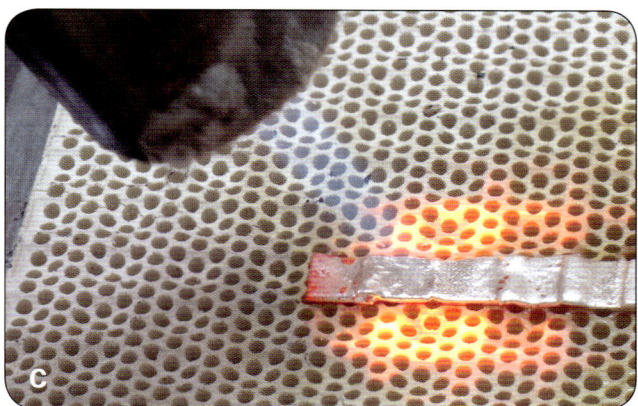

The blank for this ring can be made from scraps or strips of sheet of the same gauge. Since the ring gets harder to form after fusing, it's better to use thin gauges, like 22 or 24.

Fuse a textured blank

For a textured ring, cut some scrap 24-gauge sterling sheet into similar size pieces. They don't have to be square, but the closer they are in width, the easier it will be to lay them out like a ring blank for fusing. If you're starting with a bigger piece of sheet, set a pair of dividers to 5–8mm and scribe a parallel guide line **[A]**. Since the pieces are going to be overlapped for fusing, the thickness of the metal for the Ring Blank Size by Metal Gauge chart (p. 109) will be twice the gauge of the sheet.

For example, this ring uses 24-gauge sterling sheet, which is .5mm thick. Two layers would be 1mm thick, so I used the 18-gauge column on the chart for my size. Since the blank will be made from lots of little pieces, draw a line as long as the ring blank size with a permanent marker on the cold honeycomb block, adding about 5mm to allow for shrinkage during fusing.

Flux each piece on all sides before fusing **[B]**. Use paste flux; the wet flux will help hold the pieces together during layout. Line them up along the guideline, overlapping like a line of cards. If they overlap in the same way, by at least 1mm, it's easier to make into a ring later. Butt joins or uneven connections are harder to fuse and break easily.

Fusion happens near the melting point of the metal. The metal will glow a bright red or light orange and the surface will look glossy and wet. Start by heating the entire blank, letting the flux settle and liquify into a clear glaze. Use the pick to realign any parts that moved. Focus the flame near one end of the blank, where the first pieces overlap. When the join is glossy on both sides at the same time, the metal has fused **[C]**. Move down the line, fusing each piece to the next. If the metal is overheated, it can shrink and melt. But a little extra heat can create more texture. After fusing the pieces together, heat them again, letting the surface texture before pulling the heat quickly away. Quench and check the joins. Pickle for 20 minutes, since the piece will be heavily scaled from the high temperature.

Fuse a nontextured blank

For a nontextured ring blank, use shears to cut two to four 3mm wide strips of 24-gauge sterling silver sheet, referencing the chart on p. 109. The strips will be overlapped to make the width of the band. Use the column for twice the thickness of 24-gauge, which is 18-gauge, to find the ring blank size, and add 5mm to allow for shrinkage on the ends. Texture the strips with

Can you really fuse sterling?

Yes! Many metals can be fused. Fine silver, which contains no copper, and Argentium sterling are predisposed to fuse easily. Other metals, like sterling, copper, and brass, need to be protected with flux during fusion, or firescale can prevent the metal from bonding. Since they are a little harder to fuse, there is more chance that the metal will begin to melt and reticulate into bonus textures.

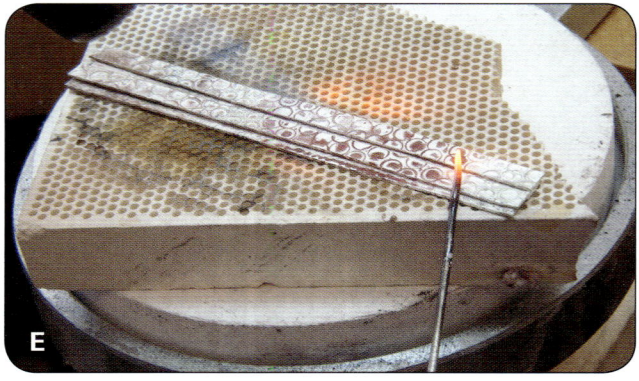

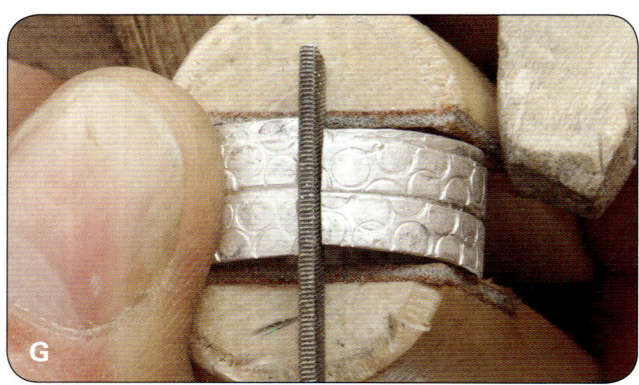

stamps or texture hammers. Contrasting textures are recommended. Straighten any curves by tapping along the convex edge with a rawhide mallet on a steel bench block **[D]**.

Flux each piece, front and back, and arrange them on the honeycomb block. The long edges should overlap for easier fusion. Start by preheating the entire blank until the flux turns clear. Focus the heat on 13mm (½") sections, watching for the silver to turn glossy as the edges fuse **[E]**. Pull the heat away just after the edges fuse for the least amount of texture.

You can press down any gaps with the pick, but be gentle: The metal is very soft during fusing and the pick can make deep marks. Keep moving down the line, fusing a little section at a time. Rotate the block and fuse again from the other side, concentrating on the joins that were hard to see from the other side. Quench in water and check your joins. If there are any gaps, close them with a mallet after pickling. Re-flux and fuse again. If you're using a honeycomb board, notice the interesting texture of dots on the back of the blank from the metal melting into the honeycomb pattern.

After pickling the blanks, straighten and trim them to size before forming the rings: File one side straight with a cut 0 hand file. Use this side as a guide to scribe a parallel line with dividers for the width of your blank. File or shear the silver to this line. Cut and file one end of the blank to 90 degrees for the first half of the join. Set the digital calipers to the length for your ring blank from the chart, and scribe the trim mark. Cut and file it to size. Both ends should be straight and 90 degrees.

Form the band around a ring mandrel. Align the joins with flat/half-round pliers **[F]**. Overlap the ends a few times to create enough tension to close the join. If there are gaps, hold the ring with a ring clamp and use a flat cut 2 needle file or saw through it a few times with a #2 blade **[G]**. Apply Cupronil on all surfaces of the ring. Solder the join with one or two 1mm balls of hard solder **[H]**. Pickle for 10 minutes.

Fuse the border rings

Reshape the fused bands on the ring mandrel with a rawhide mallet. Check the size. If they're not correct, adjust the size before making the borders (see *Adjusting Ring Size*, p. 108). The sides of the ring should be filed flat for a better fit with the two rings that will be made for the

borders. Sand them on each side with 220-grit abrasive paper until they lay flat against a steel block with no gaps.

The sample rings are tapered at the join for a more comfortable fit. To make a taper, mark an equal amount to remove from either side of the join, framing the amount of ring you want to have left on the bottom. File to that line on both sides with a flat cut 0 hand file **[I]**. File across it, filing evenly on both sides of the shank, to make an even taper back to the original width at the top **[J]**. Sand flat on 220-grit paper, to make good solder joins.

The borders are made of 14-gauge Argentium round wire, which is easy to fuse. Its smooth surface contrasts nicely with any torch texture. On the Ring Blank Size Chart, find the length for the matching ring size using the 14-gauge column. Cut two matching lengths of wire with flat, flush ends. Form the rings around the mandrel, align the joins with flat/half-round pliers, and close them with overlapping tension. Lay the ring on a clean honeycomb board, without flux, and heat the join until the metal glows red and the join glosses over and fuses **[K]**. This process will be much faster and easier than the sterling!

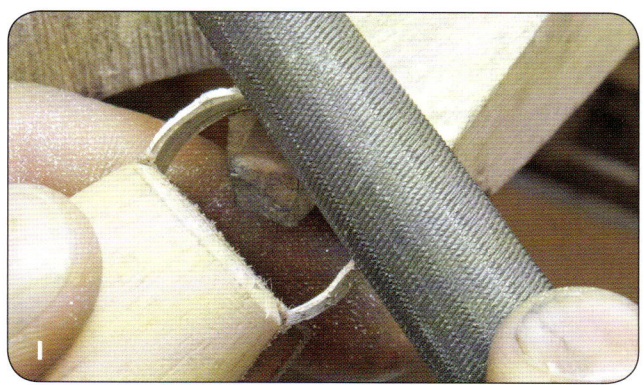

What is Argentium sterling?

Argentium is 930 parts fine silver. Some of the copper that is part of sterling silver alloy is replaced with germanium in Argentium sterling, making it resistant to tarnish and firescale. It fuses like fine silver and hardens like sterling. In fact, when heated without flux, the germanium forms an oxide that actually can clean the Argentium, leaving it almost matte white. Flux inhibits this cleaning scale, so no flux is used when fusing and only on the join during soldering. Argentium has a lower melting point (starting at around 1410°F/766°C), so avoid using hard sterling solder which flows at around the same temperature.

Sounds like a great metal; what's the downside? Well, Argentium is fragile, even when it's barely red hot, and slumps easily if not supported. If it's moved during this time, it can crack or split, but often the breaks are so clean that you can put the pieces back together and fuse it again. And it can be contaminated by sharing tools with other metals, which can smear their alloys on Argentium, causing it to tarnish or scale. This means keeping separate files, solder boards, sandpaper, and polishing tools for Argentium. If you mix metals, the Argentium will be a little contaminated anyway, but the surfaces can be cleaned during polishing.

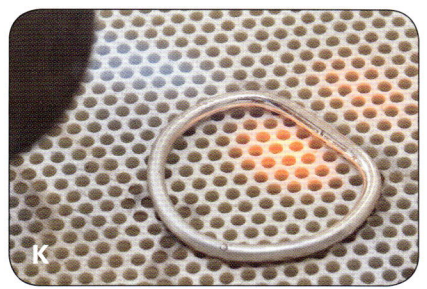

Argentium fuses so cleanly that the join line may not disappear entirely, but it will still be fused. Before turning off the torch, run the flame over any gray areas to bring up the white germanium oxide; heat actually cleans this metal. And since no flux was used, usually the metal can be quenched and then immediately formed. Just let it cool for 10 seconds before quenching or it might crack. Reshape the ring on the mandrel and check the fit against the fused blank. The rings should match the sides. If not, they can be easily stretched to fit or even cut open and re-fused to make them smaller.

Solder the rings together

Sand one side of each Argentium ring on 220-grit paper to make a flat surface. Flux just the flat side of each Argentium ring, and place them face up on the honeycomb board. Place a 1mm chip of medium solder every 6mm (¼") on the flat side. Heat the rings until the solder flows into a flat puddle. Flux the fused band completely with Cupronil, being careful not to get any on the Argentium. The Cupronil will protect the sterling from scale, but the Argentium needs flux only along the join because of its natural resistance. Stack the rings together with the solder sides facing the wide band. Heat evenly until the solder flows again,

joining the rings **[L]**. Remember: Argentium melts at a much lower temperature than sterling silver, so pull the heat back if it turns glossy. Let it cool for 10 seconds to avoid cracking, and then quench. Pickle for 15 minutes. Check the joins. Any tiny gaps can be closed with a rawhide mallet and soldered again.

Polish and add patina

Polish the ring using the abrasive attachments recommended below. Paint any areas you wish to darken with Silver Black on a cotton swab. Neutralize in water with a little baking soda. Remove the excess and restore the highlights using radial bristle disks.

POLISHING ABRASIVE	WHERE TO POLISH
Silicone polishing wheel, black	The joins and any unwanted texture.
Radial bristle disks: red, blue, peach, light green	The whole ring, inside and out. The flexible disks will preserve textures. If the Argentium bands need work, use cut 2 or 4 needle files to restore their shape, sand to 600-grit, and then polish with radial bristle disks.

FUSED BALL PEARL EARRINGS

These lightweight modern earrings feature a little trick: fusing the ends of wire into balls as they pass through a delicate pearl. It's not magic—it's a heat shield! But the heat required to ball the wire before it burns the pearl requires the intensity of a small oxygen/propane torch, with flames that approach 5000°F (2760°C). If the bead is more durable, like stone or glass, it's possible to use a micro butane torch.

LEVEL
Beginner

TECHNIQUES
Using heat shields

Cutting tubing with a tube cutter

MATERIALS
- ½" (13mm) 8.7mm ID sterling silver seamless tubing
- 2" (51mm) 18-gauge Argentium sterling round wire, dead soft
- **2** 2mm ID sterling 20-gauge round jump rings
- **2** 8mm round pearls
- Hard silver solder wire
- Pair of earring wires

TORCH
Small torch with #3 or 5 tip and medium flame

FLUX
Cupronil

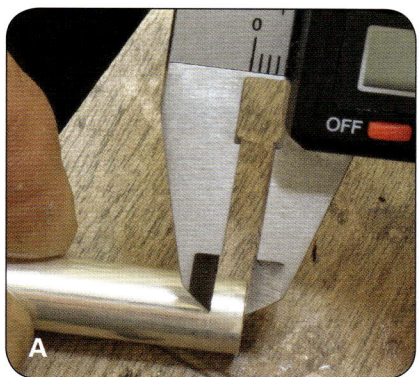

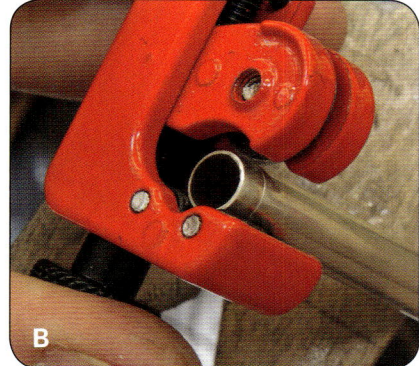

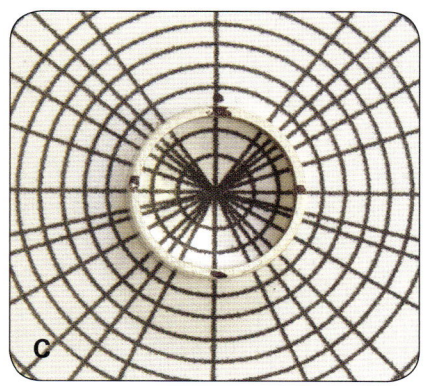

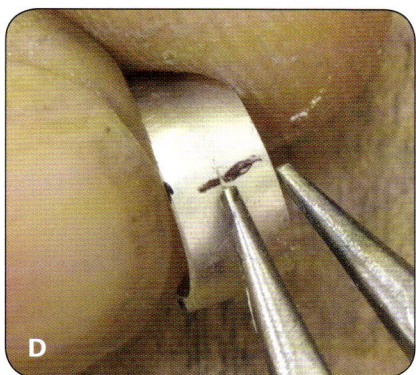

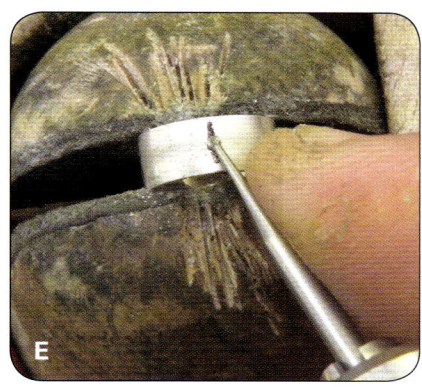

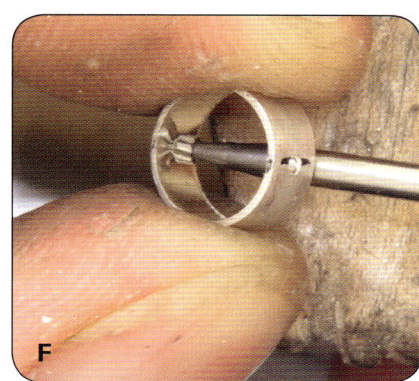

Make the tube earrings

Scribe a 4mm line on 8.75mm ID sterling tubing **[A]**. With a small pipe cutter (available at hardware stores), cut the tubing **[B]**: Clamp the sharp wheel on the scribed line and turn the cutter once around the tube. The rollers keep it aligned during cutting. Tighten the pressure knob another quarter turn. Spin the cutter around again. Repeat until the piece of tubing comes off. Repeat for the second piece. Sand the ends on 320- and 600-grit paper to remove the beveled edges.

You'll drill two holes in the tubing for the wire to pass through the pearl, and solder a jump ring on top. This takes careful alignment. Use a circle divider (see p. 110) to mark four symmetrically spaced points on each tube **[C]**. On one opposing pair of marks, scribe a centerline with dividers **[D]**. Hold the tube with a ring clamp.

Make a small divot for the drill with a 1mm ball bur **[E]**. Drill a 1mm hole on two sides. Remove any burs by hand with a 2mm setting bur **[F]**. Check the alignment of the pearl and tube with the 18-gauge sterling wire. The pearl and wire should look centered. (The next step explains how to enlarge the pearl hole.)

Scribe a centerline on the top mark for aligning the jump ring. Solder a 2mm 20-gauge sterling jump ring closed with 1mm ball of hard solder. File a small flat spot on the join. Apply Cupronil to the entire tube and set up the pieces with two third hands. Solder the jump ring to the tube with another 1mm ball of hard solder **[G]**. Hold them as far from the join as possible to limit the heat-sinking qualities of the steel. Pickle for 5 minutes. Polish the pieces using the abrasive attachments recommended below **[H]**.

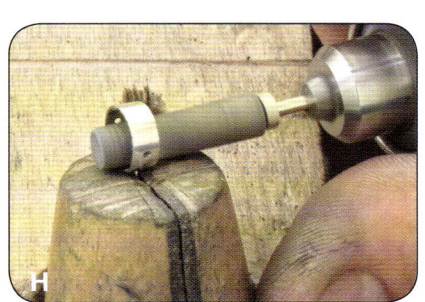

POLISHING ABRASIVE	WHERE TO POLISH
Silicone polishing wheel, black	Any excess solder.
Silicone cylinders: blue, pink	Polish inside the tubes by holding each piece by the ring with a ring clamp (they can get very hot). If there are too many scratches, back up to black and then use blue and pink.
Radial bristle disks: blue, peach, light green	The outside of the tubing and ring.

Fuse the balls

Pearls have small holes, usually as fine as 24- or 26-gauge. That would make a very small, weak join, even for earrings, so enlarge the holes with a 1mm drill. Hold the pearl with the soft leather jaws of a ring clamp, to protect yourself and because the pearl gets hot during drilling **[I]**. Drill halfway from one side, then reverse it and drill from the other side for an even hole. If you drill at medium speed, the pearl won't need to be cooled with water while you work.

I used Argentium sterling for the balls because it melts without flux, and for its faster, lower melting point. Cut two 1" (25.5mm) pieces of 18-gauge Argentium round wire. Hold one end with tweezers and melt the other end of each wire into a ball.

Here's a simple trick for estimating how much wire is required to make a ball: Cut an exact amount of wire (say, 1"/25.5mm). Melt one end into a ball. Measure what's left, from the top of the ball to the end of the wire. The amount missing is roughly how much you need to make the same size ball.

Insert the wire through the top hole and trim the other end to 6mm (¼") past the tubing. Hold it by the jump ring with a third hand. Use a toothpick to pack heat shield all around the pearl. Press it in so that it insulates the inside, between the pearl and the silver, especially on the side facing the flame. Leave no part of the pearl exposed. Clean off any heat shield on the wire or near where it will ball up against the earring.

With an oxidizing flame for the most heat, bring the tip of the cone near the end of the wire **[J]**. Keep the flame parallel or away from the

tubing—pointing it toward the tubing will burn the pearl! The wire will ball up in just a few seconds. Follow it with the flame until it is tight against the hole.

Let the earrings air cool and then scrub off any heat shield with a toothbrush and soapy water. Rinse thoroughly. There should only be minor discoloration that can be removed with radial bristle disks. Add an earring wire to each earring.

Adjusting the setting

The wet heat shield will steam, but if you see smoke or smell burning pearl, that's not a good sign. Usually this means too much heat was used for too long and too near the earring, or there wasn't enough heat shield to cover the pearl. If it's damaged, drill a new pearl and try again. If the ball won't melt or won't melt completely up to the earring, either the wire was too short, or there wasn't enough heat. A short wire is hard to get started and will burn the pearl by the time it does. It's easier to heat the end of a longer piece, at least 6mm. If the flame is adjusted too small, it may take too long to ball up the wire, or not at all. Or the hot spot at the tip of the cone may be too far away from the wire, or not aimed correctly, partially missing it.

GOLD-FILLED TUBE-SET RINGS

Gold-filled metal is an affordable alternative to karat gold for jewelry. The surface layer of gold is thick enough to withstand soldering, but too delicate for any filing or rough finishing. If the gold is stripped away, the base metal core will be exposed and will tarnish to a dark contrast. But with a little careful fabrication and finishing, you can have the luster and warmth of gold for a fraction of the price. The project steps show how to make the ring on the right; the design on the left is made in a similar way.

LEVEL
Intermediate

TECHNIQUES
Soldering filled metals

Making a tube setting
from tubing

MATERIALS
- 12" (30.5cm) 14-gauge 14K/20 gold-filled round wire, dead soft
- 3mm 3.5mm sterling silver heavy-wall tubing
- ½" (13mm) 14-gauge Argentium or sterling silver round wire, dead soft
- 3mm faceted gemstone
- Easy silver solder wire
- 10k or 14k yellow medium solder
- 4" (10.2cm) 8-gauge brass rod

TORCH
Large-flame butane torch, medium flame, or small torch with #5 –7 tip, medium neutral flame

FLUX
Paste flux

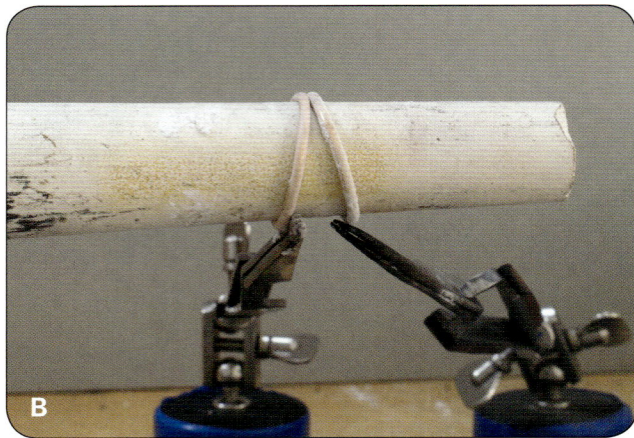

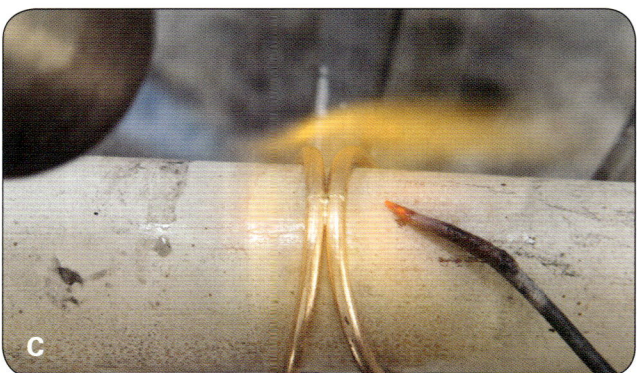

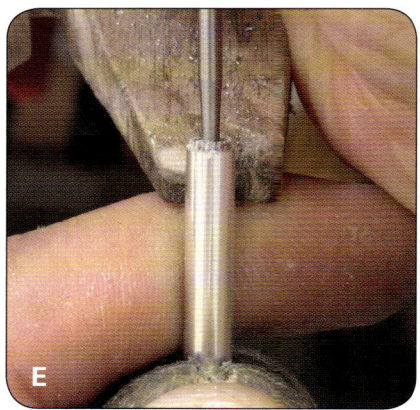

Solder the gold-filled ring

This ring is made of two rings soldered together at the base. The top is flared to fit a tube setting and a pair of silver disks. First, make the rings from 14-gauge 14K/20 gold-filled wire (see the first step of *Crown Prong Rings*, p. 74). Use the 14-gauge column on the Rings Size by Metal Gauge chart, p. 109. After forming the ring around the mandrel, take special care to align the join and make it flush. The better the join, the less trouble it will be to polish later. Apply Cupronil to the entire ring and work on charcoal or honeycomb. Solder the ring closed with a 1mm ball of 10k or 14k yellow medium solder **[A]**. Pickle for 5 minutes and reshape the rings on the mandrel. Check and adjust the size (see *Adjusting Ring Size*, p. 108). If the soldering went well, the rings should still have polished surfaces with no signs of red scale.

The rings have to be set up for the next join with the bottom of the shanks touching and the joins lined up at the bottom. This can be done with two third hands or the help of a ring stand. Ring stands are graphite or ceramic, both slight heat sinks, so use a large-flame butane torch. Apply Cupronil to both rings and place them on the ring stand, using two third hands to spread them apart **[B]**. About 5mm of metal should touch between the rings for a good join. Heat the rings evenly, and then solder with a 1mm ball of medium 14k yellow gold solder **[C]**. If the piece gets stuck on the stand, lightly heat the rings and move them down the taper with tweezers, or let it air cool and then pull them off by hand. Pickle for 5 minutes and then check the size on the ring mandrel.

Make the tube setting

Tube settings can be purchased in sterling and gold, but the choice of sizes is especially limited for sterling. Making one from scratch can teach you a lot about this kind of setting (see *Tube-setting*, p. 34).

File flat the end of a length of 3.5mm OD heavy-wall sterling tubing. A tube-cutting jig makes it easier to judge the quality of the 90-degree angle: Remove the stop, and hold the tubing in place with the lowest edge flush with top of the jig. File off anything above the steel with the flat side of a cut 2 needle file **[D]**.

Use a 3mm setting bur to cut the seat. Lubricate the bur by running it against bur lube. Hold the tube with a ring clamp and line up the bur for a straight cut. At low speed, with pressure, cut into the tubing to a depth of about one-third of the straight side of the bur. Rotate the tubing 180 degrees to compensate for any angles, and cut again, down to half of the side of the bur **[E]**. Brush out any dust and check the fit of the stone. Either place the stone table-side down on the pin and press the setting down onto it to snap it in place, or insert it from above with tweezers or a toothpick coated with sticky wax.

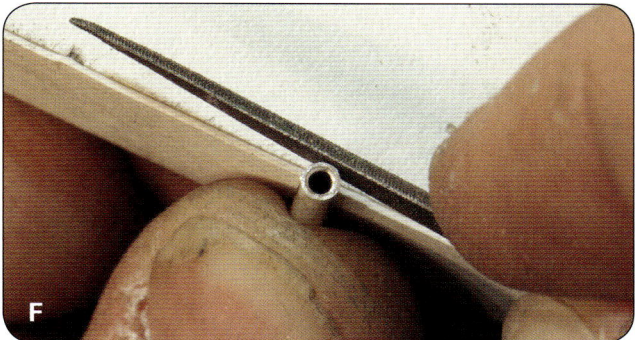

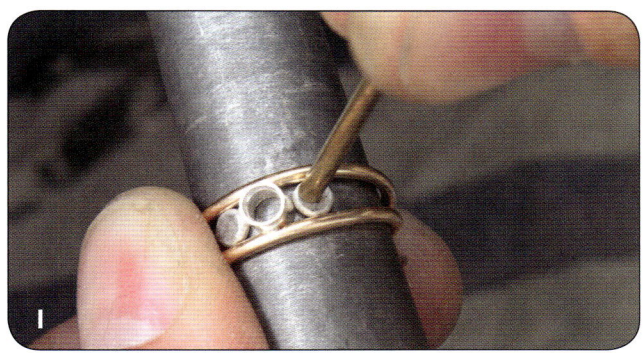

Check the depth of the bezel with a loupe. The table should be flush with the top of the bezel or slightly higher, but you should see a wall of bezel just above the crown—enough to set the stone later and burnish around it.

After the setting is the right size, check the thickness of the bezel. A thick bezel is hard to push down against the stone. If it looks too thick, file around the outside of the tubing to thin it **[F]**. Don't bevel the top edge; be sure to evenly file the entire bezel with the file flat against the side of the tube. Sand off any file marks with 320- and 600-grit sanding sticks.

Trim the setting to size with a #2 saw blade (see *Pedestal Prong Pendant*, p. 65). A tube-cutting jig can help keep the blade aligned for a straight cut. The length for the setting is 3mm. It needs to be tall enough to keep the culet inside and high enough to avoid soldering the bezel walls to the sides of the ring. Remove any burrs with a scraper and file the bottom if it's crooked.

Adjusting the setting

There are two ways to adjust the setting to fit the stone: depth and width. If the girdle of the stone doesn't fit into the setting, or sits at an angle, it's too wide for the seat. You can confirm this by removing the stone and looking at the seat with a loupe. If there are two angles clearly cut, a vertical wall and the pavilion shelf, the problem is the width. If there are no vertical bezel walls, you haven't cut deeply enough into the tubing. Adjust the width by inserting the bur, keeping it at the same level, and press it lightly outwards. Don't cut any deeper into the setting. It only takes a little bit of work at low speed, pressing in four directions, to enlarge it. Check it with the stone often, so that it doesn't become too loose.

If the girdle sits inside the setting, but there is not enough vertical bezel wall, cut the setting a little deeper. If the table sits below the level of the bezel, then it's too deep. Check the cut of the seat. If it's just a little too far, file the bezel down to the right size, being careful not to angle it. If the bur went too deep, trim the tubing with the saw and start again.

Solder the setting and disks to the ring

You will center the setting at the top of the ring, and then you will solder a silver disk on each side. For a better join with the setting, use a 3.5mm setting bur to grind the inside of the ring for a custom fit **[G]**. Apply Cupronil to the setting and ring, and set them up on the ring stand. Place a 1mm ball of easy sterling solder on the join and heat the entire ring until the solder flows **[H]**. Repeat on the other side. Try to heat slowly enough to observe which direction the solder moves. Adjust the angle of the torch to draw it into the join and away from the gold. Pickle for 5 minutes.

Melting two 6mm pieces of 14-gauge sterling silver or Argentium to make two disks (see *Mixed Metal Earrings*, p. 38). Argentium has the advantage of balling up easily and without any firescale. Pickle and sand off any scale before forging the balls flat with the polished flat face of a goldsmith's hammer. Check the fit with the ring often. The disks should snap in place for a snug join. Sand the top and bottom of each disk to a 600-grit finish before soldering. Place the ring on a ring mandrel and press them in with a pusher made from 8-gauge brass rod, if needed **[I]**. Apply Cupronil to the inside and outside of the ring and set it up on the ring stand. Solder all four joins where it touches the rings with

.5mm balls of easy sterling solder **[J]** (see *Splitting solder*, p. 29). Pickle for 10 minutes.

Polish

It's safer to burnish the ring in stainless steel shot in a rotary tumbler than to polish it with abrasives. Polish or burnish any scratches or joins before tumbling **[K]**, because the ring should be tumbled for just 30–60 minutes; longer might harden or damage the tube setting. After tumbling, a brighter shine can be made with the light green radial bristle disks. If you don't have a tumbler, carefully use the abrasive attachments recommended below.

Set the stone

The ring should be on a firm surface, like a ring mandrel, during setting. Protect the inside of the ring from scratches with painters tape around the mandrel. Insert the stone and use a loupe to confirm that it is level and at the right height. A tube setting is set like a small bezel. Keep the stone in place with a fingernail as the first four points are crimped with a prong pusher. Inspect the stone with a loupe it to make sure there are no gaps above the crown. Use a burnisher to smooth and polish the bezel **[L]**. Inspect the setting again. If there are any work marks or stubborn lumps, polish them with a blue silicone wheel and burnish again.

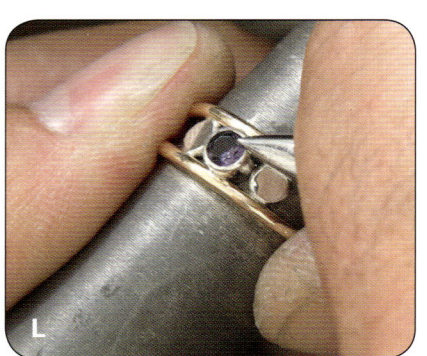

Tips for working with filled metals

- Keep filled metals protected during storage. Store wires and sheets separately in plastic bags to minimize scratches.
- Rest metal on soft leather to minimize work marks while filing.
- Joins must be aligned perfectly for soldering. There's too little gold to file or even for a black silicone wheel. Just a little filing can expose the base metal core, but you won't know until days later when it tarnishes.
- Try to match the color of the joins by using 14k or 10k yellow gold solder. A yellow tinted sterling solder is also available, but it only comes in one temperature.
- Fill joins without overfilling. Use small pieces of solder to avoid lumps.

- Prevent firescale with Cupronil or Firescoff spray flux.
- Because of the bond between the two metals, the melting point is lower. Never use hard solder. Use medium, easy, or extra-easy. A low-temperature flame such as butane is less likely to burn it. If a small torch is used, use a medium flame and be gentle with the heat.
- Try to fix any scratches or problems with burnishing, which will polish without abrasion. A tumbler with mixed stainless steel shot is a safe way to polish (see *Burnishing with a Tumbler*, p. 106).
- Use only the finest grits for polishing: blue silicone wheels and blue through light-green radial bristle disks.

POLISHING ABRASIVE	WHERE TO POLISH
Burnishers	Rest the ring on leather and burnish any join lines in the rings, rubbing the metal over the line to hide it. Also burnish any scratches. Filing or sanding will expose the base metal.
Radial bristle disks: blue, peach, light green	Polish the inside and outside of the ring with peach and light green radial bristle disks. If there is any scale, sweep just those areas with the blue disks.
Silicone polishing wheel, blue	To troubleshoot any excess solder. Blend back to a polish with light green radial bristle disks.

STERLING AND GOLD TUBE-SET BIRD PENDANT

Sculpting sterling silver into figurative shapes, such as this bird, leaves, and branch, is a nod to the beautiful jewelry of the Arts and Crafts Movement and Art Nouveau in the early 20th century.

Adding a little bit of gold, even just a few settings, adds value to your jewelry. Soldering gold and sterling together isn't difficult, and the yellow adds warmth to the green peridots. Making this pendant, however, will test your fabrication skills to successfully solder this very clean and polished design.

LEVEL
Intermediate

TECHNIQUES
Tube-setting

Sculpting metal into figurative shapes

Soldering gold and sterling silver

MATERIALS
- 2x2" (51x51mm) 16-gauge sterling silver sheet, dead soft
- 2x½" (51x13mm) 20-gauge sterling silver sheet, dead soft
- 1½x1½" (38x38mm) 24-gauge sterling silver sheet, dead soft
- ½" (38mm) 4mm OD sterling silver seamless tubing
- **2** 2mm 14k yellow gold tube settings
- 3mm 14k yellow gold tube setting
- **2** 2mm faceted gemstones
- 3mm faceted gemstone
- Easy, medium, and hard silver solder

TORCH
Large-flame butane torch, medium flame, or small torch with #5–7 tip and a melting tip, medium flame

FLUX
Cupronil

89

templates, actual size

Saw, file, and sculpt

Trace or photocopy the templates and glue them to the sterling silver with rubber cement: Glue the frame to the 16-gauge sheet the branch to the 20-gauge, and the rest of the pieces to the 24-gauge sheet. Let the glue cure for 15 minutes before sawing with #2 blades for the 16- and 20-gauge, and a 2/0 blade for the thin 24-gauge sheet. Saw just outside the lines to preserve them for filing.

Use digital calipers to check the symmetry of the circle frame and make any corrections. The top of the frame is beveled, leaving the inside edge the original 1.3mm thickness. Bevel it first with the flat side of a cut 0 hand file, then use 320- and 600-grit sanding sticks to make a satin finish. Sand the bottom flat, through 600-grit, with the paper taped to a smooth, flat surface.

Use split mandrels, with 320- and 600-grit strips, to sand the inside edge of the frame, being careful not undercut the circular shape. For a final touch, burnish along the top corner of the inside edge, making a 1mm bevel with a mirror polish.

Check the fit of the branch with the frame and mark where to solder the two ends **[A]**. Scribe a line where the frame overlaps the branch. Trim it to size and file the ends to match the inside curve. Use the half-round side of a cut 2 needle file to bevel the sides of the branch **[B]**. Make it undulate like a real branch, accenting where one branch forks into two. Use a round mini file to bevel the inside of these narrow forks, adding more dimension to what was once just flat sheet. Polish it with polishing pins: black, brown, and green **[C]**.

Hold the leaves with a ring clamp and clean up the edges with a half-round needle file **[D]**. Leave the paper on as a guide for stamping the veins of the leaves with a line stamp **[E]**. These are easy stamps to make from a flat head screw driver or by filing the end of a mild steel punch from the hardware store. Overlap the stamp marks to chase smooth lines. Remove the paper and planish the surface with rounded hammer marks from a ball peen hammer. Hold the small leaf with your fingernail or with something like a wooden clay tool, available at art stores **[F]**. Bevel the edges with a mini half-round file to frame the texture. Polish the bevels with pin polishers to a satin finish.

The bird is made from two pierced pieces of 24-gauge sheet, one for the body and one for the wing. Use two cupped nail sets from the

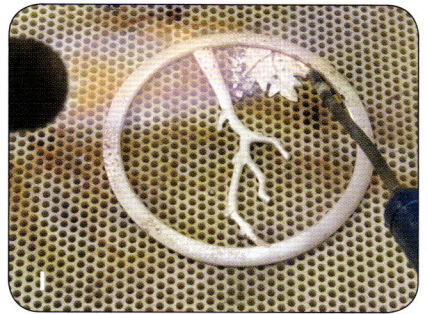

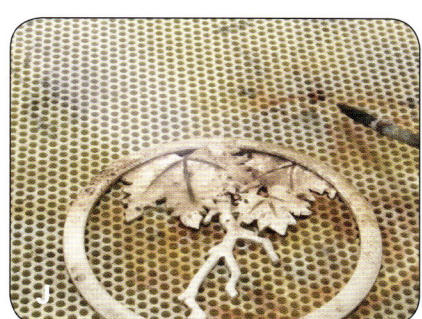

hardware store, one large (4–5mm) and one small (1.2mm), to stamp the eye on the bird. Practice first on scrap metal. Stamp the small round eye first. Tilt the larger stamp back a little and stamp a partial circle for the top of the eye. It will be sculpted with simple bevels after sweat soldering on the wing with two or three 2mm chips of hard solder **[G]**. Use Cupronil to prevent firescale. Pickle it for 5 minutes, and then file the edges to remove any join lines. File from the outside to bevel and soften the bird. Use the corner of a half-round needle file to file a groove into the wing at the notch, making it look like another level of feathers. Round the top edge of the wing a little, where it runs along the body of the bird, but leave it thick enough to be shadowed by patina later. Use black and blue silicone wheels, with their edges shaped to 45 degrees against a separation disk, to polish off any file marks and leave a smooth, fine finish. The sharp angle will make it easier to get close to the edges without damaging them, and the flat side of the wheel will make it easier to sand without rippling the surface.

Arrange the pieces together inside the frame to adjust any joins for soldering **[H]**. Overlap the frame over the edges of the leaves and scribe lines to match the inside curve. File to the line for a tight join. You may have to modify the fit again during soldering if pieces shift.

Solder the pieces inside the frame

The parts are soldered in one at time, starting with the lowest level. Apply Cupronil to all sides of the frame and branch. Solder them together, using the marker lines as guides, with 1mm balls of hard solder. Keep the heat back and work the feathered tip of the flame over the surface, getting closer as you direct the solder into the join. Make sure the joins are complete. Flux and place the lowest leaf against the inside of the frame. Solder it with two 1mm balls of hard solder **[I]**. Let the piece air cool and fit the next two leaves in place, curving the large top leaf, if necessary. Reapply some Cupronil to the piece in progress, and flux the new leaves. Solder the large leaf where it touches the inside of the frame, the lower leaf, and along the branch with small .5 mm balls of medium solder (see *Splitting solder*, p. 29) **[J]**. Do the same with the last leaf. It only takes a little solder; don't overfill these joins or they'll spill over the texture. Pickle for 5 minutes before attaching the bird and settings. Check the joins to make sure that all of the pieces are connected.

The bird is sweat-soldered in place with medium solder. Place a 1mm ball on each of the branches, in the center of where they will be under the bird, and melt them into little puddles. Heat the entire pendant and bird until it settles as the solder flows **[K]**. Test the join by placing .5mm balls of solder one at a time, along the

edges that touch the branch. When they flow it's a good sign that the bird has also completely sweat-soldered.

Carefully apply Cupronil to the tube settings, trying to coat the insides as well. Solder them on one at a time with .5mm pieces of medium solder **[L]**. Heat the entire pendant, and then brush the area around the settings with the feathered tip of the flame. Heat mostly the silver, being careful not to overheat the setting, which could cover it with solder or melt it. Watch out for the tips of the branches, which are also easy to melt. Using too much solder can flow up the sides of the setting, freezing part of the bezel. Pickle for 10 minutes and inspect all the joins with a loupe. Make sure the solder has run down along the entire length of the branch under the bird. Repeat for additional settings.

Attach a bail

Curve the 4mm OD tube bail with nylon-jaw bracelet-forming pliers (see *Pedestal Prong Pendant*, p. 65). Place it on the back of the pendant and mark where to trim **[M]**.

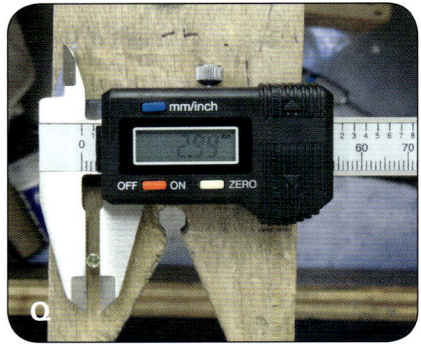

Saw it to size with a #2 blade. File the ends flat and sand to a 600-grit finish with sanding sticks. File a flat edge for soldering along the bottom of the bail with a cut 0 flat hand file **[N]**. Apply Cupronil to the entire pendant and the tube bail. Protect the previous joins with yellow ochre **[O]**. Melt three 1mm balls of easy solder where the tubing will rest against the pendant. Place the bail solder-side down on the pendant. Heat the entire pendant and bail evenly until the solder flows and the bail settles against the back **[P]**. Quench. Remove any yellow ochre with a soft brush and soapy water. Pickle for 20 minutes.

Polish
Most of the piece has been polished during fabrication. If the Cupronil protected the metal, only the last stages of polishing will be needed. Use the abrasive attachments recommended below for a final polish.

 tip

Work over a sweeps drawer or towel to catch the stones if they fall.

Set the stones
Measure the size of the stones with digital calipers **[Q]**. They should be a close match to the setting size and burs. Rest the pendant on a firm surface, like a steel block, but protect it with a piece of leather. Check the fit of the 3mm stone with the setting, inserting it with a toothpick and sticky wax **[R]**. If the stone is too big, use a 3mm setting bur to enlarge it, being careful not to bur too deeply or to angle the seat. Crimp the sides in with a prong pusher, as you would a bezel **[S]**. Burnish the edges to remove any lumps and small gaps. Repeat to set the 2mm stones. Check all the settings with a loupe and correct any gaps.

POLISHING ABRASIVE	WHERE TO POLISH
Radial bristle disks: blue, peach, light green	The entire piece. Use stacks of six disks. Don't wear down the bezels on the tubes. Use stacks of three of each grit to get into narrow areas.
Silicone polishing wheel, black, and polishing pins: gray, brown, green	Touch up any excess solder or scratches. Use pins to polish any hard-to-reach areas, and blend those areas back up to a mirror finish with radial bristle disks.

FLUSH-SET GEMS CUFF

A pattern of brilliant, flush-set stones accents this simple but elegant sterling silver cuff. The cuff is formed around an oval bracelet mandrel and planished with polished hammers to harden it into a durable, long-lasting piece of jewelry.

LEVEL
Intermediate

TECHNIQUES
Flush-setting

Planishing and burnishing

MATERIALS
- 6x1" (152x25.5mm) 14-gauge sterling sheet, dead soft
- 4" (10.2cm) 8-gauge brass rod
- **5** 2mm round faceted stones
- **2** 3mm round faceted stones

TORCH
Large-flame butane torch, maximum flame, or small torch with melting tip, medium flame

FLUX
Cupronil

Form, texture, and polish the cuff

Determine your cuff size (see *Nouveau Western Bezel Cuff*, p. 59). Trim the sheet to the blank length with a saw and #2 blade. File the ends flat with a cut 0 flat hand file. The side to be textured should also be sanded to 600-grit to remove any scratches that might show through the hammered finish. Use a split mandrel and the flex shaft to speed up sanding.

Texture one side with the slightly domed face of a polished planishing hammer on a smooth, bracelet-size steel block **[A]**. If the blank curves slightly, place it edgewise on the block, convex edge up, and hammer it back into a straight line. Anneal the blank on a charcoal block, protecting the silver from scale with Cupronil flux on all sides. Pickle for 10 minutes.

Straighten the blank with a mallet. Sand off any scale, especially on the hammered side, with radial bristle disks, white through blue. Round the ends of the blank for comfort. Saw the corners off at matching 45-degree angles. Then round them over with a cut 0 flat hand file. File the long sides to flat, straight edges. File all edges with a cut 4 hand file, and then continue with sanding sticks, through 600-grit **[B]**.

Form the blank over a bracelet mandrel starting with the ends **[C]**. Rest the mandrel on a sandbag. Adjust the shape with nylon-jaw bracelet-forming pliers, if necessary **[D]**. Using the bracelet mandrel as an anvil, planish the cuff again with the same hammer to restore the mirror finish and harden it. Planishing with a polished hammer can remove light scratches, so look closely at the surface as you work, and repeat planishing where necessary.

Move the cuff to a hard plastic block and planish the edges so they have a matching texture **[E]**.

Polish the cuff before setting the stones so they aren't damaged by the abrasives. Follow the recommendations for abrasive attachments on p. 95.

When planishing, make sure the metal is resting on the steel where your hammer strikes. If the metal is unsupported, the hammer blow will sound dull and the strike will distort the cuff.

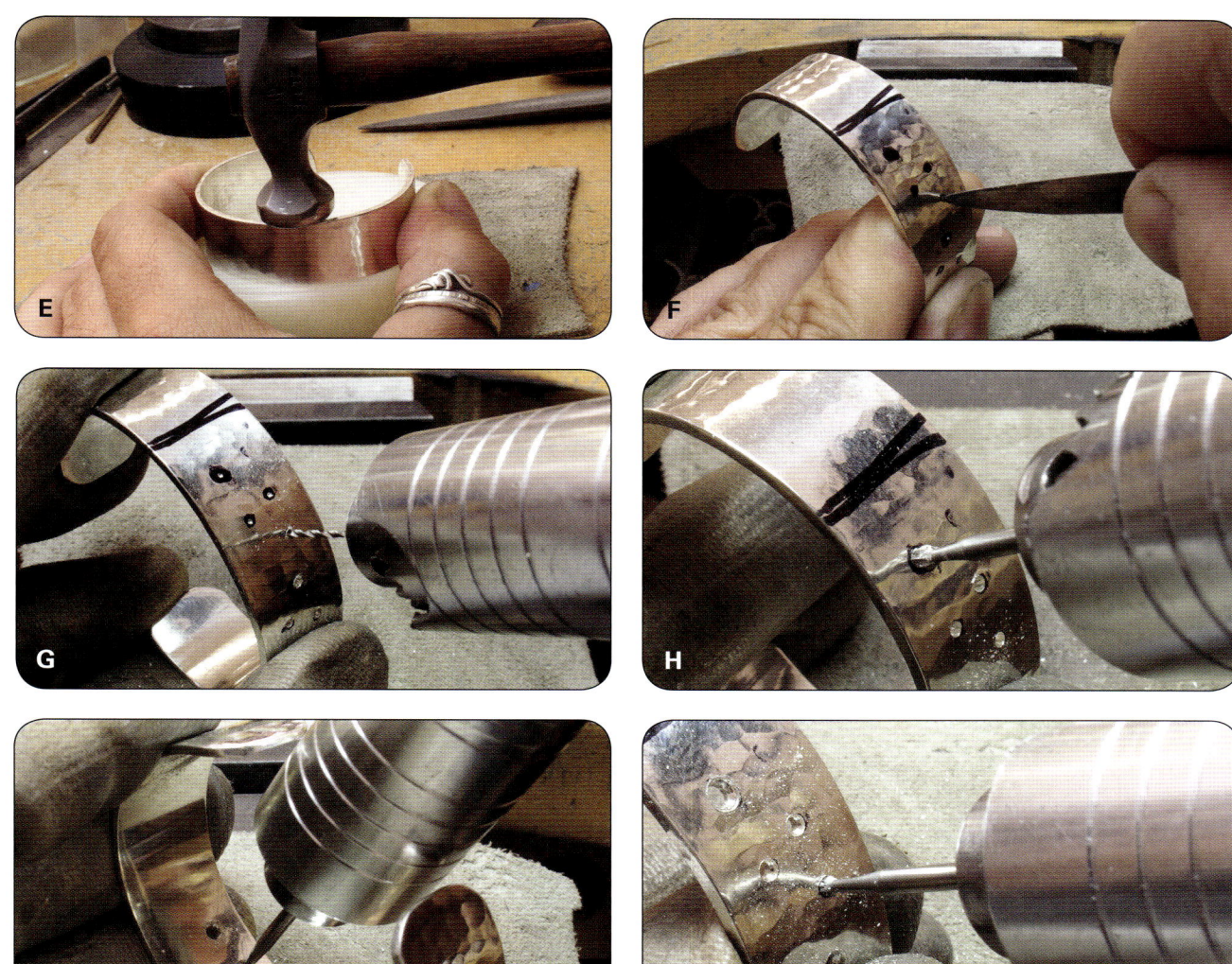

Set the stones

A flush setting is made in four steps: Drill a hole, expand with a bud bur, make a seat with a setting bur, and burnish (see *Flush-setting*, p. 35). Mark the positions for the stones with dots, spacing them 5–10mm apart. Make a divot in the center of each dot with the point of a scraper [F]. The point will dig easily into the silver, and the divot will help keep the drill bit in place. Drill at each divot with a 1mm drill bit [G]. Keep the drill bit perpendicular to the flat surface where the setting will be to avoid making a crooked setting. And keep your fingers way from the inside of the cuff or you'll drill yourself! Use less pressure as you near the end of each hole, or the chuck will mar the surface of the cuff. Expand the holes for the 3mm stones with a 1.6mm drill bit.

Next, widen each hole all the way through with its respective bud bur: 1.6mm for the 2mm settings, and 2.3mm for the 3mm stones [H].

Use low speed and cut lube with the drills and all burs to avoid overheating and dulling them. Remove burrs around the edges of the holes on the inside of the cuff by making a slight bevel with a 3mm setting bur [I]. The setting bur will cut the seat for the stone to the right depth. The bur should be the same size as the stone for a tight fit. Keep the bur perpendicular to the surface and bur down until you start to form a vertical wall [J].

POLISHING ABRASIVE	WHERE TO POLISH
Split mandrels: 320- through 1200-grit	Inside the cuff to remove any reverse marks from the steel block.
Radial bristle disks: peach through light green	Inside and outside the cuff to a mirror finish. If you have a bench-top polisher, use 2" (51mm) radial bristle disks for faster polishing. Don't forget the edges!

Place the stone with some sticky wax on a toothpick **[K]**. Press the stone in place with a pusher made from 8-gauge brass rod **[L]**. The table should be flush with the surface. Continue burring and checking the fit until the stone clicks in place at the right depth. If the girdle won't fit into the setting, then the stone is too big or the bur is too small. (See *Gold-Filled Tube-Set Rings*, p. 85, for tips on how to widen the setting.)

The stones are set by burnishing a rim around the stone to go over the girdle, holding it in place. Start the setting by burnishing four opposite points, as you would for a bezel, with a needle burnisher **[M]**. Hold the burnisher at a 45-degree angle, with the point touching the crown. Make sure the stone doesn't tilt as you burnish each point. This should help keep the stone in place for the rest of the setting.

Next, burnish the areas between each point at the same angle. Raise the burnisher to vertical, and glide it along the crown as you press outward against the rim to finish burnishing the metal down and over the girdle **[N]**. Check the fit with a loupe and by pressing the stone up from underneath with the point of the burnisher. If the stone is loose, it won't take much pressure to pop it out. Burnish until the stone is secure.

 tip

The soft brass of the stone-pusher won't harm most stones.

Troubleshooting

A flush setting requires precise fitting and careful burnishing. If the seat is too big, it will be hard to set the stone. Similarly, if the stone is too shallow in the setting, with too much of the crown above the surface, there may not be enough metal to burnish over the girdle. Another common problem is slipping with the burnisher and scratching the metal. A lot of these problems can be fixed with more burnishing.

If the stone is loose, burnish the flat surface around the edges of the setting, moving the metal inward toward the stones. This will close the setting and trap the stone **[O]**. Repeat around the inside of the setting with the needle burnisher to make a nice rim. Similarly, scratches are best burnished out and then blended back into the rest of the finish with the same polishing bits.

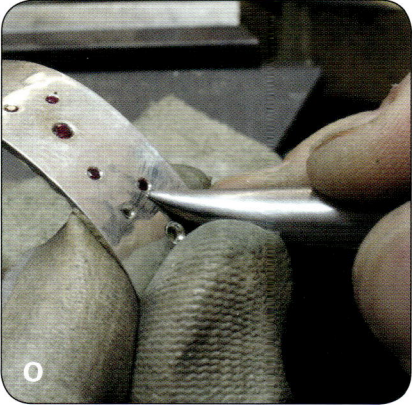

GOLD GALLERY WIRE BEZEL RING

The first time you work with gold can be intimidating, but it's not all that different from silver or copper. The biggest fear is that the gold will be ruined, but it's actually easy to recover and turn in to a metal refiner for credit toward new metal.

There are many karats and colors of gold, with their own quirks to learn. White gold is very rigid. Rose gold scales easily. 14k yellow gold is malleable. Each variety needs its own matching color and karat of solder. Gold is a little less prone to scale than sterling, but it should still be protected with a fire coat of Cupronil or alcohol and borax. Otherwise, it's a delight to work with and you can get used to it very quickly.

LEVEL
Intermediate

TECHNIQUES
Soldering gold

Using alcohol and flux

Making a gallery-wire bezel

MATERIALS
- 3" (76mm) 20-gauge square 14k yellow gold wire, dead soft
- 3" (76mm) 14k yellow gold gallery wire
- 6" 15.2cm) 14-gauge 14k yellow gold half-round wire
- 13mm round cabochon
- 14k yellow gold hard, medium, and easy solder

TORCH
Micro butane torch, maximum flame, or small torch with #5–7 tips, medium neutral flame

FLUX
Alcohol and boric acid

Self-pickling flux with a squeeze bottle and needle spout

A

bezel is too small, it can be gently stretched with a rawhide mallet on a bezel mandrel, striking the bottom edge of the gallery wire downward with the corner of the mallet to enlarge it.

The bottom of the bezel is open, but soldering on a sheet metal base is expensive. Instead, save money and increase the light available to the stone by soldering a ledge on the inside. Form 20-gauge square 14k yellow gold wire into a ring that fits snug to the inside bottom of the bezel. It's easier to curve the wire to match around a bezel mandrel or with flat/half-round pliers. Solder it with a 1mm ball of hard 14KY solder **[F]**. Reshape it on the bezel mandrel and fit it to the bezel for soldering. It should be tight enough not to fall when the bezel is turned upside down. Solder it with three or four 1mm balls of 14KY medium solder **[G]**. If you have a tendency to overflow the solder, paint the pattern of the gallery wire with yellow ochre gouache anti-flux after applying the flux. Pickle for 5 minutes and check the fit again with the stone. Sand the bottom of the bezel flat with 320- and 600-grit paper. Don't sand away the details on the bottom of the gallery wire.

Form the bezel

Start with a 90-degree flush-cut on the gallery wire. Make the cut in an easy place to hide a join. Form the gallery wire around the girdle of the stone and mark where to trim it **[A]**. Cut the bezel in a matching place to continue the pattern. Close the bezel with flat/half-round pliers, being careful not to crush the pattern. File the join, if needed, with a flat needle file.

B

Any soldering on gold starts with dipping the piece into the boric acid fire coat and stirring it around. Place the piece on the charcoal and ignite it with the flame. The alcohol will burn off with a green flame and leave a flat white coat of borax **[B]**. Drip a separate, self-pickling flux on the join and heat to dry it in place **[C]**. Solder the bezel upside down to limit the chances of melting the fine tips of the gallery wire **[D]**. Use a .5mm ball of hard 14k yellow gold solder. Pickle for 5 minutes and then check the fit, looking at the flat bottom of the stone **[E]**.

The stone should pass easily through the setting from top to bottom, without any gaps. If the

Solder on the ring shank

The ring shank is a split-style, made from two pieces of 14-gauge 14k yellow gold half-round wire. The length for the blank is a little more interesting. Because the shank is open underneath the setting, it's a little shorter. Find the measurement in the 14-gauge column of the Ring Blank Size by Metal Gauge chart, p. 109. Measure the diameter of the base of the setting with digital calipers, and subtract that amount,

C

D

Safety tips for alcohol flux

Many goldsmiths use denatured alcohol-based flux for gold. It's mixed with boric acid, and when ignited, the alcohol burns off, leaving a flat coat of flux. It's no secret that an alcohol-based flux can catch fire, especially with a flame around a jar of the stuff. If the flux catches fire, don't panic. Smother the flame with a towel if any spills catch fire. If a flame ignites on the solder board or similar resistant surface, just let it safely burn out. If the jar catches fire, put the lid back on to snuff out the flame. This is why alcohol flux jars often have a lid around the brush: for easily extinguishing any accidents. Obviously, keep the jar well away from the flame and make sure the brush isn't on fire before putting it back in the jar. If the whole idea of working with alcohol and fire gives you the willies, there's great news: Nonalcohol fire coats such as Cupronil can be used on gold.

in millimeters, from the length from the chart. Flush-cut the pieces to that size. Straighten the wires with a rawhide mallet and steel block. Flare the ends away from each other by bending them with chainnose pliers. At least a third of the length should touch for soldering. Pin them down to a soft charcoal block with 19-gauge dark annealed binding wire. Brush the wires

with alcohol flux and ignite. Add self-pickling flux to the join. Solder with two 1mm balls of hard 14KY gold **[H]**. Let it air cool and then remove the pins. Pickle for 5 minutes.

Form the shank around a ring mandrel at the correct size. Place the shank upside down on the back of the setting and mark any place to trim it

to center it without anything extending into the opening. File flat spots on top of each wire so it rests flat on the setting for soldering **[I]**.

The shank is sweat-soldered in place on a heat-reflective surface such as charcoal block or honeycomb board. Coat the pieces in alcohol flux. If you're concerned about the ring falling

apart during soldering, paint the join of the shank with anti-flux. Place a drop of self-pickling flux on the flat spot on each wire. Melt a .5mm ball of easy 14KY solder on each one. Center the ring on the setting. Heat both parts evenly until the solder flows between them **[J]**. Pickle for 5 minutes and inspect all of the joins for any gaps.

Polish and add patina

Before setting the stone, polish the ring using the abrasive attachments recommended below. Gold can be antiqued with Silver Black, but only if a steel brush is used to apply it. The reaction creates fumes, so use ventilation and wear a mask. Antique just the gallery wire and the bottom of the shank. Neutralize it in cold water and baking soda.

Set the stone

The little prongs on the gallery wire are easy to set, but the stone may be too low. Test the fit. If it is low, raise it up with another ring of 20-gauge 14KY gold square wire **[K]**. Form it as in the first step. It doesn't have to be soldered to the setting. Polish it before putting it in place. In fact, you may have to press it in with an 8-gauge brass rod as a pusher **[L]**. Hold the ring on a ring mandrel. Clean the stone and insert it into the setting. Push the prongs on the gallery wire down in pairs of opposites with a burnisher **[M]**. Polish off any work marks.

POLISHING ABRASIVE	WHERE TO POLISH
Silicone polishing wheel: black	Remove excess solder and join lines. Be very delicate with the gallery wire. If the join was well aligned for soldering, there should be very little to remove.
Silicone cylinders: black, blue, pink	Inside the setting. Use black to remove excess solder or the join line, and blue and pink to polish it. Shape them to flat top cylinders on separation disks to make it easier to polish inside.
Pin polishers: gray, brown, green	Inside the notches of the shank.
Radial bristle disks: blue, peach, light green	The entire piece, inside and out.

3 Basics Review

ANNEALING

Metal is annealed to a dead soft, more malleable state by heating it, usually to a dull red glow, and then quenching it. Sterling silver and gold are annealed at around 1200°F (649°C), when they glow a light red color. Brass and copper anneal at a medium red color. As soon as the redness disappears, quench them in water.

1. Flux. Metals like sterling, copper, and brass will get lots of firescale if annealed without flux. Plus, flux turns clear at 1100°F (593°C), so flux protects your metal and tells you when it is starting to anneal. Flux all sides completely with a fire coat like Cupronil (see *Using spray flux*, p. 22).

2. Place on a charcoal block. Charcoal reduces firescale because it reduces oxygen during heating. Metal facing the charcoal will be the least oxidized.

3. Heat. Dim the light, if possible, to better see the color as you heat the metal evenly with a torch. Keeping the surface covered in a large flame will also reduce firescale, as the flame reduces oxygen. Use a large-flame butane torch or small torch with a melting tip. Another trick for spotting the annealing point is to mark the metal with a fine-tip permanent marker (Sharpie) before adding flux. The marker lines will disappear when the metal is near annealing temperature.

4. Hold. Back the flame away and hold a light red color on the surface for a minute. Don't let it get too hot; a bright red or orange color is a sign of impending melting or a rippled reticulated texture.

5. Quench. Let the metal cool for a few seconds until it turns dark and no longer glows red. Quench it in water. Sterling, brass, and copper are more malleable if quenched after annealing.

6. Pickle. Soak in hot pickle for 10–15 minutes. If the metal was well fluxed, it should be free of scale. If it wasn't, the surface will still be black or blotchy with coppery red or gray scale.

SAWING

Saw blades come in sizes, from coarse #4 down to ultra-fine 8/0. Match the size of the blade to the gauge of the metal; the rule of thumb is to have several teeth of the blade on the thickness of the metal. Coarse blades are more durable and better at cutting straight lines, and finer blades are easier to turn to saw out details, but they break easily.

To load the saw frame, rest the top of the frame on the table and place the blade with the teeth facing up in the top clamp. They should feel smooth if stroked toward the handle and rough when stroked away. The top end of the blade should be flush with the top of the clamp, and pointed directly across the saw to the clamp near the handle. Tighten the top clamp as tight as you can by hand. Adjust the size of the frame by loosening the back clamp, until ⅛" (3mm) of the end of the saw blade sits inside the bottom clamp, but don't clamp it. This will leave space for adding tension to the blade later. Tighten the back clamp. Press the top of the saw against a sturdy table. Lean your upper body or hip against the handle to compress the frame, moving the blade further into the bottom clamp. When the blade has moved in at least ¼" (6mm), hold the pressure on the saw frame and tighten the bottom clamp. Release the pressure.

The blade should be taut. It will "ping" sharply if plucked, and if the blade is pressed gently with your finger it will have very little flex. If it sounds dull when plucked or the blade feels loose, open the bottom clamp and press the blade in further.

Keep your fingers in sight and a safe distance from the front of the blade. If you have to hold close to the teeth, protect your fingers by curling them under, as though you are chopping vegetables with a knife.

The saw is used with the V-slot bench pin. Use your dominant hand to hold the saw, and your other hand to hold the metal down against the pin. Use 2 or 3 fingers, often in a V formation, to hold the metal against the pin to keep it from bouncing and to turn the metal as you saw. It's OK to reach through the saw frame. Lubricate the blade with paraffin or a product such as Cut Lube or Bur Life. To start, hold the blade at a slight forward angle and use short strokes to file a notch. Saw just outside the outline; you can file back to the outline later.

Keep your sawing hand relaxed. Too much pressure will slow down sawing and break the blade. Saw with the blade perpendicular to the metal. Sawing at an angle will bevel the edges and the blade will bind or break. To make sharp turns, keep the blade moving up and down in place, and slowly turn the metal to the new direction. Turning without moving the blade will snap it.

FILING

The higher the number of the cut, the finer the file. Use coarse files to remove metal quickly and fine files, up to a #4, to leave a smooth finish. Hand files are larger for faster work, and smaller versions, called needle and mini files, are for detail work. They come in different shapes, such as half-round, which is handy for filing straight edges and curves. Use the flat side for straight edges and convex curves, round side for concave.

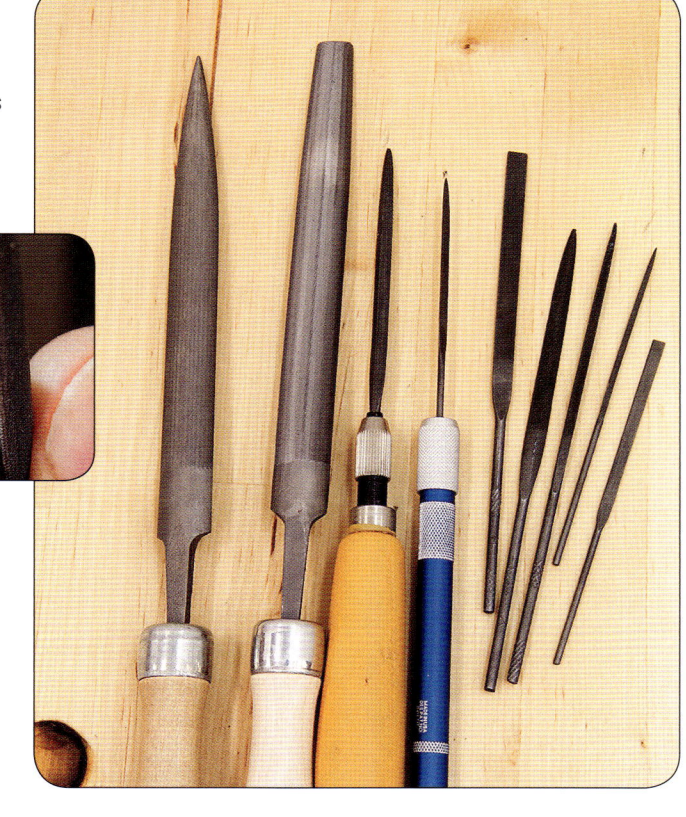

File forward on the push stroke, away from you. Filing backward will not remove much metal. Let the edge you're filing overhang the side of the pin. Filing close to the side of the pin will help you keep the file perpendicular, making straight edges on both sides of the piece. Filing at angles or working unsupported will create random, rough, distorted edges. For pierced areas, move the metal to the V-slot and use needle files.

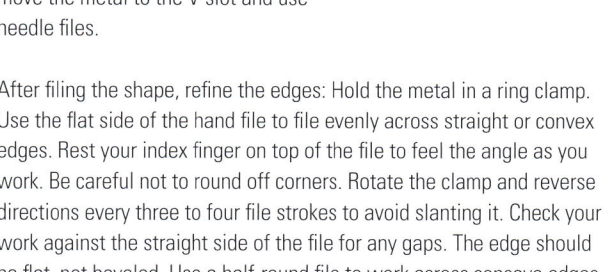

After filing the shape, refine the edges: Hold the metal in a ring clamp. Use the flat side of the hand file to file evenly across straight or convex edges. Rest your index finger on top of the file to feel the angle as you work. Be careful not to round off corners. Rotate the clamp and reverse directions every three to four file strokes to avoid slanting it. Check your work against the straight side of the file for any gaps. The edge should be flat, not beveled. Use a half-round file to work across concave edges.

SANDING

Sanding removes scratches and polishes your metal. Use abrasive paper made for metal: wet/dry sandpaper from a hardware store or micron-graded polishing paper from jewelry suppliers. The higher the number, the finer the grit. Sand in one direction for consistent sanding lines, which makes it easier to see scratches. Always sand across scratches. Change directions often to avoid slanting the edges. Sand perpendicular to the previous sanding lines when moving up to the next finer grit.

For more control, rest the paper flat on a smooth surface such as a bench block. This will make the truest, flattest surface for polishing or making

a flush solder join. For fast sanding, rest your metal on a piece of leather, fold a 1x4" (25.5x102mm) strip of sandpaper, and sand from above. The paper will conform to any slight curves in the metal.

Sanding sticks are great for sanding edges and are easy to make. Glue sandpaper with rubber cement to a straight, flat piece of wood (paint sticks work great). Brush glue on the wood and the back of the paper. Trim the paper to size after the glue cures (about 5 minutes). Label the stick with the grit of the paper.

WET OR DRY SILICON CARBIDE SANDPAPER						
	MICRON-GRADED POLISHING PAPER					
180- / 220-grit	320- / 400-grit	600-grit	1200-grit	4000-grit	6000-grit	8000-grit
COARSE ————————>		SATIN BRUSHED FINISH ————————>				MIRROR FINISH

POLISHING WITH POWER TOOLS

Tips for polishing with power tools

Work with support
To control your movements, brace your hands together and against a bench pin to support the tool and the piece as you work, or rest your hands, forearms, or elbows against the table. Don't let the jewelry touch the table, since that can cause scratches.

Use the bottom edge of the wheel, not the face
Polishing with the face or flat plane of most wheels will break them and is harder to control. Polish with the bottom edge of the wheel. Stroke the wheel in line with its edge or with the attachment at a 45-degree angle to the edge.

Start in the middle and polish off the edges
Start with the tip of the attachment in the center of the piece and pull it toward you and off the edge. If you start on the far edge, the rotation of the tool can grab the piece and flip it out of your hand.

Polish slowly with even pressure at the right motor speed
Drag the tip of the attachment slowly across the metal to give it time to do its work. Apply even pressure. Radial bristle disks should flex as you polish. Polishing takes longer and is ineffective if you press too lightly or move too quickly. Use the right motor speed for your abrasive; too much speed can damage the attachments and your jewelry.

Shape attachments against a file or separation disk
The edges of rubber wheels and polishing pins get rounded with use, which can gouge a surface. Reshape them by running them at low speed against something abrasive—an old coarse file or a separating disk.

Don't polish away texture or details
Texture and details can vanish if you polish with coarse abrasives such as silicone wheels. Use less pressure and polish for a shorter time. Use fine abrasives, like 400-grit blue radial disks, and polish in the same direction as a texture, not across it. Use polishing attachments that conform to the surface; radial bristle disks are a good choice.

Polishing abrasives
Polishing can be messy: Dust from felt and bristle wheels charged with polishing compounds like rouge can stain. My preference is to use split mandrels loaded with sandpaper and abrasives such as silicone polishing wheels, radial bristle disks, and pin polishers, which leave less dust and debris to clean up. These come in progressively finer grits that you can use to polish your jewelry to a mirror shine.

Use the chart on the next page to compare the pros and cons of each abrasive, to identify their order, and for advice on what speed to use with them. Polishing attachments are problem solvers, allowing you to remove scratches and defects with the right tool. Use the second chart to compare the coarseness of each bit and its recommended task to learn when to switch to the next abrasive.

Silicone polishing wheels or cylinders
Silicone polishers come in wheels, cylinders, and pins. Wheels are available with flat edges for flat surfaces and knife edges to get into detailed areas. Cylinders or bullets fit inside rings and small spaces. The coarse and medium wheels will remove metal and excess solder, and can polish away details (including texture). They come without mandrels and have to be mounted on mini screw mandrels. Cylinders use threaded mandrels. Pins require a special holder for use with a flex shaft.

Radial bristle disks
These flexible disks can remove firescale and polish flat surfaces, wire, and texture to a beautiful finish with no dust. Stack three to six disks on a mini screw mandrel. Looking down from the tip of the mandrel, the bristle tips should point right and the 3M stamp should face down, in the direction of the rotation of the tool. If the tips face the wrong way, they will bend and break.

Split mandrels and sandpaper
Cut a piece of sandpaper in half lengthwise and then cut 1x4¼" (25.5x108mm) strips. Mount a split mandrel in the chuck with the base as close to the jaws as possible. For a right-handed person, thread the paper into the slot with the abrasive side facing you. The short end should be on the right of the tool and curled over slightly. When used at low speed, the paper will curl around the bit. Press it against the metal to form a tight drum. Protect the hand holding the metal with cotton finger guards (less risky for injury than gloves).

Silicone polishing wheels and cylinders.

Radial bristle disks.

Polishing pins.

	SILICONE WHEELS & CYLINDERS	RADIAL BRISTLE DISKS	POLISHING PINS	SPLIT MANDRELS & SANDPAPER
PROS & CONS	**Pros:** Good for flat surfaces, removing excess solder, and scratches. **Cons:** Can remove texture and distort wire and flat surfaces.	**Pros:** Flexible bristles polish texture, wire, and sheet. **Cons:** Won't remove excess solder or reshape metal.	**Pros:** Detail hard-to-reach places, inside jump rings and bails. **Cons:** Can leave track marks from the small size of the bit.	**Pros:** Sands flat surfaces evenly. Easy to remove used paper to expose fresh abrasive. Fits inside rings and other small spaces. **Cons:** Rough on texture. Can flatten details.
SPEED	Slow	Medium	Slow	Slow

GRIND & SHAPE			
Separation disk	Silicone wheel: white/coarse	Split mandrel (180- or 220-grit)	
PRE-POLISH / REMOVE FIRESCALE OR EXCESS SOLDER			
Radial bristle disks: (yellow/80, white/120, red/220)		Split mandrel (320-grit)	
Radial bristle disks: blue/400	Silicone wheel: black/medium	Split mandrel (400- or 600-grit)	Polishing pins: gray/medium
POLISH			
Radial bristle disks: peach/6μ	Silicone wheel: blue/fine	Split mandrel (1200-grit micron-graded polishing paper)	Polishing pins: brown/fine
Radial bristle disks: light green/1μ	Silicone wheel: pink/extra-fine	Split mandrel (2500+-grit polishing paper)	Polishing pins: green/extra-fine

BURNISHING WITH A TUMBLER

Burnishing with stainless steel shot in a rotary tumbler can rub away small scratches, but it's not abrasive. It won't polish away firescale, excess solder, file marks, or deep scratches. For the best results, remove those defects first with sanding or power tools, stopping at a satin 400-grit finish, and then tumble your pieces to a bright polish.

Tumbling can damage soft stones or beads, so either don't tumble them, tumble before adding stones, or tumble for a shorter time. To use, load the tumbler barrel with two pounds of shot, a tiny drop of liquid dish soap or burnishing solution, and just enough water to cover the surface of the shot. Add your pieces and tumble for 20 minutes to 2 hours. Rinse with water and let the jewelry pieces and the shot dry.

ADDING PATINA

If you like your jewelry to be antiqued, with dark patina in the recesses and highlights buffed back to reveal the polished metal, you need to use a chemical. Silver Black can create an antique look on silver, bronze, or gold. Liver of sulfur works well on silver and copper. Other patinas for specific metals such as brass are available from jewelry suppliers. Pickling, soldering, and excessive cleaning or polishing can remove patina.

Silver Black

Silver Black, a hazardous but effective acid-based product made by Griffith, immediately turns silver black. Use Silver Black cold. Apply by dipping, or paint just the areas you want dark with a cotton swab. Neutralize the application in water with baking soda. Leftover solution can be poured back in the bottle and used again. If you want to antique gold, you must apply it with a steel brush. Steel will make a nasty vapor, so wear a respirator and work outside.

Liver of sulfur

Liver of sulfur (LOS) is a smelly mixture of potassium sulfides and is available as lumps, liquid, or gel. I recommend using it as a gel in a squirt-top bottle, which means less mess. When added to hot water, it darkens silver and copper, but doesn't work well on brass or gold. Store it in a dark place, since air and light make LOS go bad.

Boil some water separately. Never heat LOS anywhere shared with food. Pour just enough hot water into a glass or plastic container to cover your jewelry. Add a squirt of LOS gel, around the size of the tip of your finger or up to a teaspoon. Stir to mix it into a yellow-green color. Dip your metal using copper tongs or wire, or make a sieve by drilling holes in the bottom of a plastic cup. Let it soak for a minute until the metal darkens. You can also apply the solution with a natural-bristle brush, but it takes

longer to darken. Lighter colors like yellow and blue can be made with a lukewarm or a weak mix, or by removing the metal after only a few seconds. Stop the patina and neutralize it with cold water and a little baking soda.

Remove patina from the highlights

Remove the patina from the highlights with a polishing pad, or repeat the last round of polishing with a power tool. Stay out of the recesses; polish only the highlights. Leaving your piece too long in the patina will make a more durable patina, but it takes stronger polishing or sanding to bring back the highlights. You can also tumble-polish pieces after removing the patina.

Patina safety

While using any patina solution, you should wear gloves and eye protection. Dispose of excess solution according to the recommendations of the manufacturer. Never use these products in the same area where you prepare food or with any food utensils. Take precautions to contain spills: Work over a tray or inside a large, open container. Use in a ventilated area or outdoors. Keep all chemicals away from children and pets. Don't store solutions near metal, chain, or tools, because it can discolor them.

Grease and dirt will interfere with patina application, so make sure the metal is clean. When applying patina solution, have a glass or plastic cup for the solution, another cup for rinse water, and a third cup with water and baking soda to neutralize the patina. The first rinse cup will absorb some patina but won't stop the reaction.

Don't mix baking soda with your patina solution; it can ruin it. Dispose of solution and clean up any spills by neutralizing with baking soda. Liver of sulfur smells bad and can bubble over during neutralizing, so do it outdoors and inside of a larger container.

MEASURING RING SIZE

Ring sizes in the U.S. range from 1–15 and higher. To find your ring size, either use an existing ring that fits and find its size on a ring mandrel, or size your finger with a finger gauge.

A ring mandrel is a steel, wood, or plastic mandrel that is engraved with measurements for standard ring sizes. Whole sizes are marked with a wide line. A dash between two whole sizes indicates a half size. To measure a ring, place the ring on the mandrel as far as it will go. Measure the size from the center of the ring itself. For example, a size-8 ring would have the line for size 8 in the middle of the band.

A finger gauge is a set of rings calibrated to match standard sizes. Finger gauges can be plastic or metal, and thin or wide bands. A wide ring band will fit your fingers differently than a thinner ring. A thinner band can slip to the end of your finger, near the web, where it naturally tapers. A wide band fits over more of your finger and so has to be a little larger to fit comfortably.

When you use a finger gauge, the ring should turn easily on the finger and slip off with a little effort. If it's too tight to turn, it's the wrong size. If the gauge comes off too easily, it's too big.

Note: Fingers swell in the summer and shrink in the winter. A loose ring in summer will fall off in winter. A ring that's tight in winter may not come off in the summer.

ADJUSTING RING SIZE

If your ring is the wrong size, you can stretch the ring a little bigger or remove material to shrink it.

Increase up to a half-size with stretching

Pull the ring down on the mandrel as far as it will go. Using a mallet, strike the ring with a downward stroke against the top edge to force it to stretch a little. Hammer like this once around the entire ring. Remove and reverse the ring on the mandrel and repeat. Check the size and repeat as necessary.

Increase 1–2 sizes with planishing

Planishing with a steel hammer will thin the band, but it will stretch it faster and further than with a mallet. Place the ring on the mandrel as above and hammer evenly around the ring band. Keep checking the size as you work, and reverse the ring after each round. For wider bands, you should concentrate your planishing on the half of the ring that sits on the thick end of the mandrel. The other side sits above the tapering steel; hammering there will slant the band like a cone.

Increase a size or more by adding material

If your ring is too thin, delicate or detailed to stretch by planishing, or if it has to be bigger than you can safely stretch, increase the size by adding material. Use the sizing chart below to find the amount to add. Find the current size of your ring and the size you want on the chart. Set a pair of dividers to the distance between the two sizes. Use a matching shape and gauge of metal for your additional piece. Scribe the amount you need on the metal with the dividers and cut it with your pliers. Use a jeweler's saw to cut the ring open at the join. Each side should be flat for soldering. Insert the new material and pinch it flush with the edges of the join. After soldering, file away any excess metal and reshape the ring on the mandrel.

Decrease your ring size by removing material

To shrink your ring, you have to remove some of it and solder the ring closed again. To find the right amount, use the sizing chart on this page. Find the current size of your ring and the size you want. Set a pair of dividers to the distance between the two sizes. Find the join on your ring and scribe the amount to remove, starting either with the join line or with the line in the middle of the amount to remove. Use a jeweler's saw or flush cutters to remove the excess and close the join (see *Crown Prong Rings*, p. 74). Don't leave the original join elsewhere on the ring or it may fall apart during soldering.

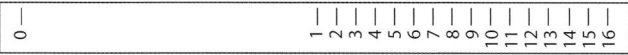

Sizing chart

RING BLANK SIZE BY METAL GAUGE
THICKNESS IN MILLIMETERS (B&S GAUGE IN PARENTHESES)

U.S. SIZE	2.1 (12)	2.0	1.8	1.6 (14)	1.4	1.3 (16)	1.2	1 (18)	.8 (20)
5	55.6	55.3	54.7	54.0	53.4	53.1	52.8	52.1	51.5
5.25	56.2	55.9	55.3	54.6	54.0	53.7	53.4	52.7	52.1
5.5	56.9	56.6	56.0	55.3	54.7	54.4	54.1	53.4	52.8
5.75	57.5	57.2	56.6	55.9	55.3	55.0	54.7	54.0	53.4
6	58.1	57.8	57.2	56.5	55.9	55.6	55.3	54.6	54.0
6.25	58.8	58.5	57.9	57.2	56.6	56.3	56.0	55.3	54.7
6.5	59.4	59.1	58.5	57.8	57.2	56.9	56.6	55.9	55.3
6.75	60.0	59.7	59.1	58.4	57.8	57.5	57.2	56.5	55.9
7	60.6	60.3	59.7	59.0	58.4	58.1	57.8	57.1	56.5
7.25	61.3	61.0	60.4	59.7	59.1	58.8	58.5	57.8	57.2
7.5	61.9	61.6	61.0	60.3	59.7	59.4	59.1	58.4	57.8
7.75	62.5	62.2	61.6	60.9	60.3	60.0	59.7	59.0	58.4
8	63.2	62.9	62.3	61.6	61.0	60.7	60.4	59.7	59.1
8.25	63.8	63.5	62.9	62.2	61.6	61.3	61.0	60.3	59.7
8.5	64.4	64.1	63.5	62.8	62.2	61.9	61.6	60.9	60.3
8.75	65.0	64.7	64.1	63.4	62.8	62.5	62.2	61.5	60.9
9	65.7	65.4	64.8	64.1	63.5	63.2	62.9	62.2	61.6
9.25	66.3	66.0	65.4	64.7	64.1	63.8	63.5	62.8	62.2
9.5	66.9	66.6	66.0	65.3	64.7	64.4	64.1	63.4	62.8
9.75	67.5	67.2	66.6	65.9	65.3	65.0	64.7	64.0	63.4
10	68.2	67.9	67.3	66.6	66.0	65.7	65.4	64.7	64.1
10.25	68.8	68.5	67.9	67.2	66.6	66.3	66.0	65.3	64.7
10.5	69.4	69.1	68.5	67.8	67.2	66.9	66.6	65.9	65.3
10.75	70.1	69.8	69.2	68.5	67.9	67.6	67.3	66.6	66.0
11	70.7	70.4	69.8	69.1	68.5	68.2	67.9	67.2	66.6
11.25	71.3	71.0	70.4	69.7	69.1	68.8	68.5	67.8	67.2
11.5	71.9	71.6	71.0	70.3	69.7	69.4	69.1	68.4	67.8
11.75	72.6	72.3	71.7	71.0	70.4	70.1	69.8	69.1	68.5
12	73.2	72.9	72.3	71.6	71.0	70.7	70.4	69.7	69.1
12.25	73.8	73.5	72.9	72.2	71.6	71.3	71.0	70.3	69.7
12.5	74.5	74.2	73.6	72.9	72.3	72.0	71.7	71.0	70.4
12.75	75.1	74.8	74.2	73.5	72.9	72.6	72.3	71.6	71.0
13	75.7	75.4	74.8	74.1	73.5	73.2	72.9	72.2	71.6
13.25	77.3	77.0	76.4	75.7	75.1	74.8	74.5	73.8	73.2
13.5	77.9	77.6	77.0	76.3	75.7	75.4	75.1	74.4	73.8
13.75	78.6	78.3	77.7	77.0	76.4	76.1	75.8	75.1	74.5
14	79.2	78.9	78.3	77.6	77.0	76.7	76.4	75.7	75.1

For rings wider than 4mm, add .5mm to the length.

CIRCLE DIVIDER

This template can help you divide a disk into even segments. (*Fused Ball Pearl Earrings*, p. 82, is one project that uses this tool.) Center the disk on the template and use a permanent marker or scribe to mark equidistant lines on the disk.

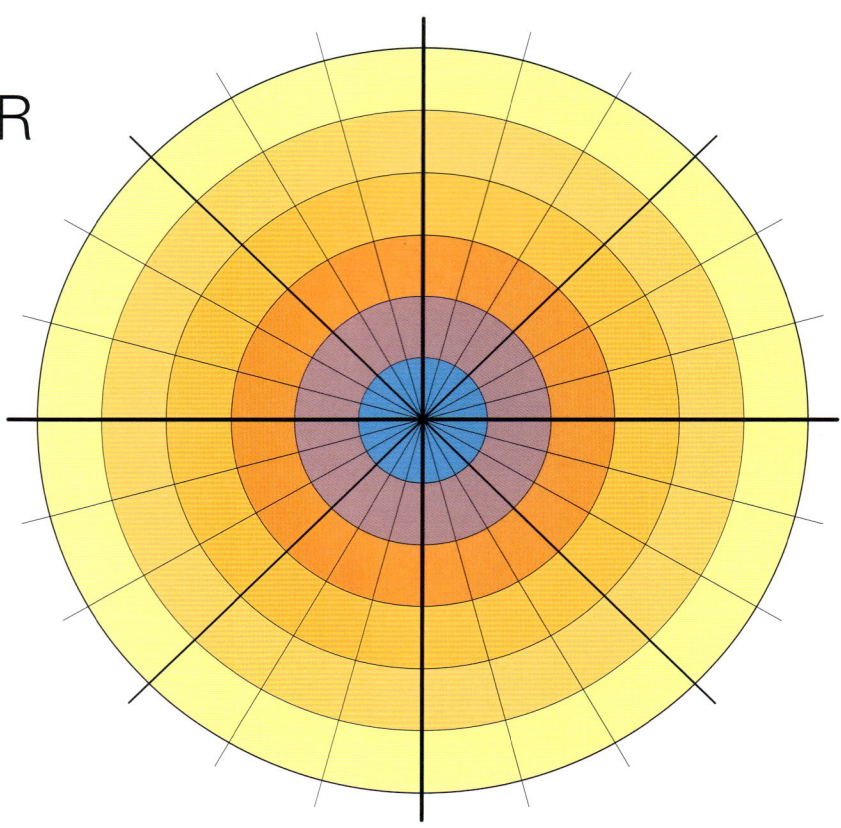

ACKNOWLEDGMENTS

Thank you to my beautiful wife, Anat, who took care of the school and made it possible for me to concentrate on my writing.

Thank you to my teachers, Al Pine and Dieter Muller-Stach. Although they have passed on, their teaching and love for the craft lives on through their students.

Thank you to my talented editor, Mary Wohlgemuth, who was my advocate and partner for this project from the very beginning.

And thank you to my students, who continue to teach me as much as I teach them.

ABOUT THE AUTHOR

Joe Silvera is the author of *Soldering Made Simple: Easy Techniques for the Kitchen-Table Jeweler*. He earned a B.F.A. degree with honors in metalsmithing, and later apprenticed as a goldsmith in southern California. After his apprenticeship, he worked as a model-maker for jewelers in Los Angeles and sold his own jewelry at fine art galleries and shows across the country. His favorite technique is lost-wax casting; he especially enjoys carving detailed sculptures of animals for rings and other jewelry.

Joe and his wife run a popular jewelry school in the San Francisco Bay area. Their classes provide a great foundation in fundamental jewelry skills, mixed with laughter, encouragement, and lots of creativity. Visit silverajewelryschool.com for updates and information about classes, workshops, and more.

INDEX

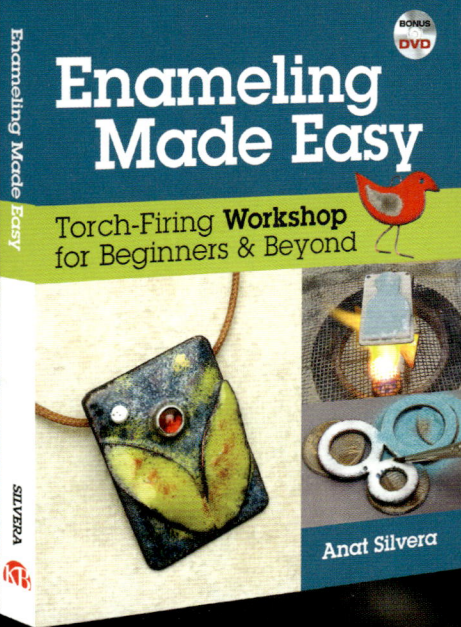